Genes, Chromosomes, and Disease

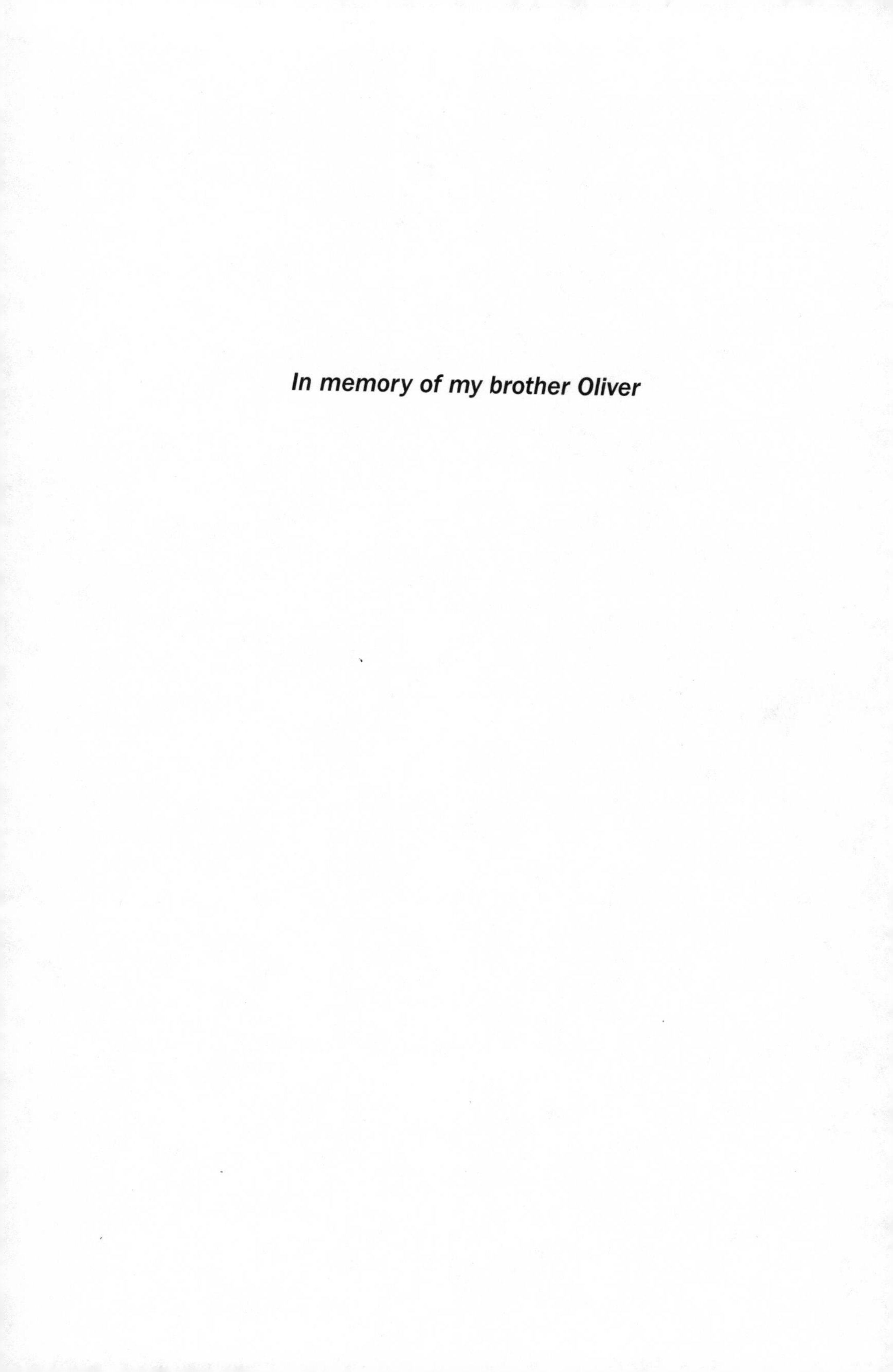

In memory of my brother Oliver

Contents

Preface

The science of genetics began in 1900 with the independent rediscovery of Mendel's 1866 paper by Carl Correns and Hugo de Vries. Until the middle of the nineteenth century, blending theories of inheritance prevailed, but it became clear to Charles Darwin and his cousin Francis Galton that the hereditary elements must be particulate to provide the kind of variation upon which natural selection could work. Each of them proposed a particulate theory of inheritance, but the particles had to be hypothetical as the architecture of the cell and its different components were only beginning to reveal themselves to the curious eye. By 1900, a great deal was known about cell structure. In particular, chromosomes had been identified and Walther Flemming, a German scientist, had characterized their behavior in cell division (mitosis). Another German scientist, Theodor Boveri, provided evidence that chromosomes of the germ cell lineage provided continuity between generations. And in 1902, an American graduate student, Walter Sutton, connected chromosomes with genes, in a classic paper. Thomas Hunt Morgan and his associates obtained experimental proof of the chromosome theory using the fruit fly *Drosophila* as a model. Working with *Drosophila* in his Fly Room at Columbia University, Morgan and his colleagues would elucidate many of the most important principles of Mendelian genetics.

In England, William Bateson became Mendel's great advocate. One would have thought such advocacy unnecessary except that, just about the time of Mendel's rediscovery, Francis Galton had come up with a model of inheritance, which he called his Ancestral Theory. Particularly in Great Britain, there was much controversy in the first decade of the twentieth century between Galton's supporters and Bateson. The Mendelians finally won out. In the course of these heated exchanges, Bateson became aware of the work of an English doctor, Archibald Garrod. Garrod was studying a disease called alkaptoneuria that caused the urine to blacken. His results suggested to Bateson that a recessive gene mutation might be involved. Bateson entered into a correspondence with Garrod, who in 1902 published a paper titled "The Incidence of Alkaptoneuria: A Study in Chemical Individuality." And with that paper, Garrod made the first connection between a human disease and a gene.

The aim of this book is to provide an overview of the relationship between genes and disease, what can be done about these diseases, and the prospects for the future as we enter the era of personalized medicine. The first three chapters deal with diseases that are simple in the sense that they result because of single gene mutations. Chapter 1, "Hunting for disease genes," considers the pedigree and its use in deciphering human genetic diseases and, at the end, the question of how many genetic diseases there are in the context of the structure of the human genome and the genes it contains. Chapter 2, "How genetic diseases arise," is about how the process of mutation gives rise to genetic defects, but also about how this same process has produced millions of tiny genomic changes called single nucleotide polymorphisms (SNPs). Most SNPs have little or no effect on the individual, but they are of major importance to those who desire to investigate genetic diseases, particularly complex ones. People with and without a genetic disease can be compared to see if any of these SNPs can be associated with specific diseases. The chapter also considers what happens when mistakes occur in partitioning chromosomes properly to sperm and eggs. Chapter 3, "Ethnicity and genetic disease," examines the reasons why some diseases are more prevalent in some races and ethnic groups than others and explains why this has nothing to do with race or ethnicity per se.

The second group of three chapters considers genetically complex diseases. Chapter 4, "Susceptibility genes and risk factors," is about genetic risk factors and diseases like type 2 diabetes, coronary disease, and asthma, where the environment also plays an important role. In each case, there are single gene mutations that can cause the disease. These disease mutations are considered in some detail as they show how certain single gene changes can lead to complex diseases. However, people with these single gene changes only represent a small fraction of those suffering from the disease. In most people who suffer from asthma, have type 2 diabetes, or are susceptible to coronary disease, there is a complex interplay between a variety of genetic risk factors and the environment. Unraveling these interactions is a work in progress.

Chapter 5, "Genes and cancer," discusses cancer, a large collection of different genetic diseases. What they all have in common is the propensity for uncontrolled growth. It has only been possible to work out the many different genetic pathways that lead to cancer because of basic research in cell biology. This has provided the necessary background information on how the normal pathways themselves are organized. The topic of cancer genetics is so vast that select examples have been chosen to illustrate several different points concerning the disease. For example, cervical cancer shows how viruses sometimes act as causative agents of cancer. The greatly increased frequency of lung cancer in recent years illustrates that decades can elapse between the exposure of a tissue or organ to carcinogens, in this case those present in cigarette smoke, and the appearance of the disease.

Like type 2 diabetes or coronary disease, schizophrenia and bipolar disease are genetically complex, as discussed in Chapter 6, "Genes and behavior." There have been many false alarms in identifying susceptibility genes for these and other behavioral conditions—the gay gene controversy comes to mind. But there have also been some notable successes. The chapter begins by recounting the history of the "warrior gene." This odd gene has been implicated in a wide variety of bad or risk-taking behaviors.

Chapter 7, "Genes and IQ: an unfinished story," deals with a subject whose relevance may not seem apparent initially. The reader may rightly ask what on earth this topic has to do with disease. The answer is that not only do quite a number of genetic diseases affect IQ, but in the first half of the last century, the presumption that "feeblemindedness" was inherited was the basis for involuntary sterilizations, particularly of women, in many states in the United States, Scandinavia, and Nazi Germany. To this day, there are those who argue that IQ differences between races and classes are largely genetic in nature and, therefore, explain certain alleged inferiorities.

For better or worse, it seems likely that IQ and related tests will be used to measure intelligence for a long time because they yield numbers and numbers are easier for most people to deal with than descriptions. Take wine, for instance. All that business about tasting like black cherries with a hint of cinnamon loses out to Robert Parker's numbering system. However, his scale is so compressed, between the high 80s and 100, that a Bordeaux wine that rates 96 can command a far greater price than one that Parker grades as 90. IQ scores, in contrast, are not compressed and follow the pleasing shape of the bell curve. Furthermore, IQ does measure something that relates to what we would call intelligence. Most would agree that the cognitive powers of children with Down syndrome are qualitatively different from those of ordinary children. This difference is captured in IQ distributions for children suffering from Down syndrome and children without this affliction. In both cases, IQs are normally distributed, but the upper end of the Down distribution overlaps with the lower end of the distribution for children who do not have the disease. However, the data on the heritability of IQ rest on shaky underpinnings. They largely depend on comparing the IQs of less than 200 pairs of identical twins reared apart and the assumption that the environments in which these twins were reared are not correlated.

Having dealt at length with genetic diseases, the next question is what to do about them. Chapter 8, "Preventing genetic disease," discusses prevention as the most desirable outcome, particularly for the most severe genetic diseases, but how do we accomplish this? Suppose a man and woman in their late thirties get married and want to have a child while it is still possible. They have a relatively high risk of giving birth to a child with Down syndrome. What should they do? A good place to begin is to initiate a discussion with a genetic counselor. Should amniocentesis or chorionic villus sampling predict the birth of a Down child, the counselor can be helpful in explaining in a nondirective way the options open to the couple. They themselves will have to decide whether the pregnancy should continue or whether to terminate it. Or suppose another couple knows that they may give birth to a Tay-Sachs child. The couple has the choice of initiating the pregnancy and aborting the fetus if it has Tay-Sachs or planning to have a healthy baby following in vitro fertilization and preimplantation genetic diagnosis. This permits the doctor to implant embryos that will not develop into Tay-Sachs babies although some of them may be carriers of the mutant gene. The procedure is not fail-safe, however, and multiple rounds of in vitro fertilization may be required. Moreover, these procedures are costly and the couple may have ethical or religious reasons for not opting either for abortion or in vitro fertilization.

Specific treatments need to be devised for each genetic ailment and many such diseases are not treatable, as explained in Chapter 9, "Treating genetic disease." The first line of defense for diseases like phenylketoneuria is newborn screening. If left untreated, the disease causes a rapid loss of cognition and a precipitous drop in IQ. Fortunately, if a phenylketoneuric infant is given a special diet shortly after birth, these cognitive declines can be avoided. All the states have mandatory newborn screening for this disease and many others where early intervention can make all the difference. Treatment of some genetic diseases involves administering an enzyme that is missing because of the genetic defect. This sort of therapy is often very expensive and it must be continued for life.

Then there is gene therapy. After 20 years of trying, it is fair to say that, despite all the hype that accompanied gene therapy, particularly in the beginning, gene therapy has delivered very little except in the case of a couple of diseases where the immune system has been rendered non-functional. In these cases, insertion of a copy of the normal gene into certain bone marrow stem cells has proven effective. We hope that this heralds the beginning of a new era for gene therapy, possibly in combination with stem cells, a topic that is hardly discussed in this book. The main reason that this book has practically nothing to say about embryonic or adult stem cells is that, despite very encouraging results with mouse models, we have no idea how this technology is going to play out in humans. In fact, the first approved clinical trial got under way late in 2010. We hope that the disappointments that have plagued gene therapy will not also arise in the case of stem cells, but only time will tell.

Today, drugs are being developed to target specific mutational defects for cystic fibrosis and other genetic diseases, as described in Chapter 10, "The dawn of personalized medicine." It has also become clear that certain drugs are effective with people with one genetic background, but not another. Gene testing companies are measuring genetic risk for complex diseases like type 2 diabetes, and genome sequencing will soon cost around $1,000, making it affordable for a lot of people. With regard to their own genomes, the problem for most people will be an overload of information. What are they to do with it? How are they to weigh it? How much do they really want to know? We have entered the era of personalized medicine, an era in which most of us are going to need some guidance. Before proceeding to discuss the array of topics that are the subject of this book, a word about the diverse ways in which human genetic diseases are named is in order.

Genetic diseases are named in various ways. Most commonly, they bear the names of their discoverers. Down syndrome, for instance, is named for its discoverer, John Langdon Down, a nineteenth-century British physician. Sometimes the name is descriptive—sickle cell anemia comes to mind. The red blood cells of people with this disease do sickle. Sometimes the names are misleading or hard to understand. Why would anyone name a disease that can cause profuse bleeding hemophilia? Only Count Dracula would appreciate that. Or thalassemia. What's that about? It's a disease like sickle cell disease, but its name refers to the sea in Greek. The reason for this odd name is that this disease was once prevalent around the rim of the Mediterranean. Sometimes diseases are named quite specifically for the function they perturb. G6PD refers to a common alteration that results in a deficiency of the enzyme glucose-6-phosphate dehydrogenase.

1

Hunting for disease genes

Leopold George Duncan Albert, Duke of Albany, eighth child and youngest son of Queen Victoria, was buried on Saturday, April 12, 1884, in the Albert Memorial Chapel, Windsor Castle.[1] He was only 31. Leopold's pregnant wife Princess Helene, the daughter of George Victor, reigning Prince of Waldeck-Pyrmont, arrived by carriage to view her husband's remains and to shed some tears over them. Next, the Seaforth Highlanders, in which Leopold was an honorary colonel, arrived. They were wearing their medals and sidearms. The Coldstream Guards followed the Seaforths led by their band. The servants of the late Prince Albert, the servants of the Queen, and then the gentlemen of Leopold's household followed them. The coffin was borne by eight Seaforth Highlanders and followed by the Prince of Wales in the uniform of a field marshal.

Also marching in the funeral procession was a French general who had accompanied Leopold's remains from Cannes, where he had died. On March 27, Leopold had slipped on a tiled floor in the Yacht Club and injured his knee. Although it has been claimed that Prince Leopold died from the effects of the morphine he had been given to ease the pain on top of the claret he had consumed with his dinner, it seems more likely that he died of a cerebral hemorrhage.[2] Leopold was the first victim of what has been called the "Royal Disease" or hemophilia.

Hemophilia A and B, recessive, sex-linked diseases, are normally expressed only in males because a male has a single X chromosome, whereas a female has two, one usually having the normal gene. That is, women are carriers who do not show any symptoms of hemophilia. Hemophilia spread from the British royal line into the Russian,

Prussian, and Spanish royal lines through intermarriage. Its source was Queen Victoria. She had two daughters who were carriers in addition to Leopold, but her other five offspring did not express or transmit the defective gene to their progeny.

Although it is remotely possible that a hemophilia mutation occurred in Queen Victoria very early in egg formation, it is much more likely that Queen Victoria was a carrier of the hemophilia mutation because three of her children had the hemophilia gene. If the Queen was a carrier, the egg from which she arose would either have had to be fertilized by a mutant sperm from her father Edward, Duke of Kent, or else her father was not the duke. After spending many years in Europe in the company of various mistresses, notably Adelaide Dubus and Julie St. Laurent, Edward married Victoire (or Victoria) of Saxe-Coburg-Saalfeld, the widow of the Prince of Leningen, in 1818. Victoria was born the next year and Edward died in 1820.

Perhaps the sperm that fertilized the egg that produced Queen Victoria possessed the hemophilia mutation. If so, the mutation would have arisen during spermatogenesis in the duke as there is no prior evidence of hemophilia in the royal line. In his book *The Victorians,* A. N. Wilson proposes a different theory. Another man may have fathered Victoria. Wilson supposes that man may have been her mother's secretary Sir John Conroy, a man Queen Victoria detested. Conroy and Victoire were widely suspected of being lovers, but there is no evidence he had hemophilia. Even if Conroy was not Queen Victoria's father, Wilson writes, "it seems overwhelmingly probable that Victoire, uncertain of her husband's potency or fertility, took a lover to determine that the Coburg dynasty would eventually take over the throne of England."[3] If so, the presumptive interloper would have needed to work quickly. After all, Victoria of Saxe-Coburg-Saalfeld and Edward, Duke of Kent, were married on May 29, 1818, and Queen Victoria was born just a year later.

It has long been assumed that hemophilia A rather than hemophilia B was the disease transmitted by Queen Victoria because hemophilia A accounts for 85% of all cases and hemophilia B for about 14% with various other clotting defects accounting for the remaining 1%.[4] However, we now know that Queen Victoria carried a hemophilia B mutation. This finding emerges from some remarkable detective work involving the remains of the murdered family of Nicholas II, the last Russian czar.

On July 16, 1918, the czar, his family, the royal physician, and three servants were herded into the cellar of Ipatiev House in Yekaterinburg where they were held prisoner and shot by a firing squad.[5] The bodies were to be thrown down a mine shaft, but the truck that carried them began to have engine problems so the murderers dug a shallow pit as a grave, poured sulfuric acid on the bodies to impede their identification, covered the bodies, and drove the truck back and forth over the grave site to flatten it. Half a year later, a Russian investigator, Nicholas Sokolov, retrieved some valuable objects from the likely tomb, but reported no evidence of skeletal remains. He concluded that the bodies had been destroyed, but in April 1989, a filmmaker named Geli Ryabov claimed that the bodies had not been destroyed, but that they were located five miles from the site discovered by Sokolov. Ryabov and a geologist colleague had worked out the actual burial place from photographs and the original report written by the head executioner.

DNA analysis confirmed the presence of the skeletal remains of nine people. They included the czar, the czarina, three of their five children, the royal physician, and three servants. However, two of the children were missing. This was in accord with the executioner's report that he had burned two of the bodies, one of which belonged to the czar's only son Alexei, a hemophiliac. Burned bone fragments from two skeletons were found in 2007 in another grave at the site of a bonfire in the same area.[6] The fragments proved to be what was left of Alexei and his sister Alexandra.

The hemophilia A and B genes are called *F8* and *F9* because they encode clotting factors 8 and 9, respectively. DNA analysis of the F8 and F9 genes recovered from the remains revealed that only the latter gene was altered and that Alexandra was a carrier, whereas Alexei's single X chromosome had, of course, the hemophilia mutation.[7]

The pedigree of the "Royal Disease" illustrates how useful a good lineage is in attributing a specific disease to a defective gene. This chapter considers two different approaches to identifying disease and susceptibility genes. The first is to target a specific gene. The example given here is the discovery of the gene whose alteration results in Huntington's chorea. The pedigree that provided the answer was found on the shores of Lake Maracaibo in Venezuela. Once an approximate chromosomal location had been established for the gene, the investigators, led by James Gusella at Harvard and Nancy Wexler at Columbia, had to inch along the chromosome to the actual gene using various molecular techniques, a method referred to as "positional cloning" (see Glossary).

The second approach is to search for a variety of deleterious genes in a specific sect or group that exhibits characteristics such as originating from a small founding group, inbreeding, or a high incidence of several different disease genes. The Amish, Ashkenazi Jews, and French Canadians are examples. This is one approach favored by many gene-hunting companies. Once again, pedigree analysis and positional cloning play key roles.

Until the last ten years or so, these were the two major approaches to gene identification, but with the discovery that the human genome is riddled with small genetic differences called single nucleotide polymorphisms or SNPs (see Glossary and Chapter 2, "How genetic diseases arise") coupled with the publication of the human genome sequence, two other approaches became popular that do not require information from pedigrees. In the first, called the candidate gene method, the investigator makes an educated guess at a gene or genes mutation of which might lead to a specific genetic disability. The gene and surrounding DNA are compared between people with and without the condition to see whether there are any alterations specific to people having the disease. The second method is completely unbiased and involves comparing entire genomes between the two groups for differences in SNPs. These genome-wide association studies (GWAS) have the potential for discovering differences related to genes that might not normally have been suspected of causing the disease. These methods, especially the latter, are particularly well adapted to finding genetic factors underlying complex genetic diseases like type 2 diabetes (see Chapter 4, "Susceptibility genes and risk factors," for a fuller discussion). However, as the price of whole genome sequencing continues to drop rapidly, whole genome sequencing comparisons will probably replace the candidate gene and GWAS approaches.

Venezuelan adventures: the isolation of the Huntington's gene

One day in 1858, George Huntington, a boy of eight, was riding with his father George Lee Huntington, a physician. His father was making his medical rounds on a wooded road between the towns of Amagansett and Easthampton on the South Fork of Long Island when "we suddenly came upon a mother and a daughter, both bowing, twisting, grimacing. I stared in wonderment, almost in fear. What could it mean?"[8] Thus was George Huntington introduced to the disease that would later bear his name, Huntington's chorea. Huntington's grandfather, a physician like his father, migrated to the eastern end of Long Island from Connecticut in 1797. Both his grandfather and father had observed the "slow onset and gradual development" of this hereditary disease and how some of its victims "worked on their trades long after the choreic movements had developed, but gradually succumbed to the inevitable, becoming more and more helpless as time advanced, and often mind and body failed at an even pace."

Like his father and grandfather before him, George Huntington became a doctor after obtaining his medical degree at Columbia University in 1871. That same year, he moved to Pomeroy, Ohio, to set up a family practice. On February 15, 1872, he traveled five miles across the icy landscape to Middleport, Ohio, to deliver a paper to the Meigs and Mason Academy of Medicine. The academy's membership was made up of physicians from two sparsely populated counties of the same name. In his report titled "On Chorea," Huntington began with a general review pointing out that "chorea" was a disease of the nervous system whose name derived from "the dancing propensities of those who are affected by it." He noted that chorea was principally a disease of childhood. In contrast, "hereditary chorea" as he called it was confined to the few families he had observed in Easthampton as "an heirloom from generations away back in the dim past" and it did not manifest itself until "*adult* or *middle* life."

Huntington's presentation was well received, so he submitted the manuscript to the editors of the *Medical and Surgical Reporter* of Philadelphia, where it was published on April 13, 1872.[9] Huntington's paper describing what he called "hereditary chorea" was short, clear, and concise and was widely discussed, abstracted for international yearbooks, and published in its entirety in various texts. In 1915, Charles Benedict Davenport, Director of the Eugenics Records Office at Cold Spring Harbor, New York, and a member of the National Academy of Sciences, published a paper on Huntington's chorea in the first volume of its *Proceedings*.[10] Pedigree data from four families suggested strongly that a dominant gene mutation was responsible for the disease, a hypothesis that has proved to be correct.

The discovery of the defective gene that causes Huntington's chorea really begins with the folk singer and songwriter Woody Guthrie.[11] In 1956, he was arrested in New Jersey for "wandering aimlessly," a charge often brought against the mentally ill or confused. He was committed to the Greystone Park Psychiatric Hospital in Morris Plains, New Jersey, a sprawling complex of 43 buildings that opened in 1876. He remained there until 1961 by which time his condition had worsened and he was transferred to the Creedmore facility on Long Island, where he died in 1967. Following Guthrie's death, his widow formed the Committee to Combat Huntington's Disease. Milton Wexler, a doctor, joined Guthrie in her quest. His wife and three brothers-in-law were suffering from the disease.[12] Wexler's daughter Nancy was in graduate school when her mother was diagnosed with Huntington's disease. She realized she had a 50% chance of having the Huntington's mutation herself. In her PhD dissertation at the University of Michigan in clinical psychology, she explored the cognitive and emotional consequences of being at risk for Huntington's disease. She has never revealed publicly whether she has been tested for the gene. However, it seems unlikely that she will become a disease victim because she is now over 60 and has not expressed its symptoms.

Nancy Wexler was determined to try to identify the Huntington's gene. Her big break came in 1972 when Dr. Americo Negrette, a Venezuelan physician, presented a paper at a conference in the United States.[13] Dr. Negrette had set up his practice in 1952 in a remote community near the great saltwater gulf called Lake Maracaibo. He soon noticed that certain individuals were stumbling, weaving, and falling down, and concluded they were probably drunk. He learned from the residents, however, that they were not drunk, but suffered from a disease that was locally called El Mal. He soon realized that they were expressing the symptoms of Huntington's disease and published a book on the subject in 1955.

Dr. Negrette had already begun to construct a pedigree for Huntington's disease in the Lake Maracaibo population when Nancy Wexler and her team joined him 1979. Members of the relevant families live in three villages on the shores of the lake. The scientists succeeded in tracing the pedigree back to a woman named Maria Concepcion who lived in the early 1800s and had ten children. She may not have had the disease herself. It is likely that the children of hers who suffered from the disease may have inherited it from their father, possibly a sailor from Europe. By 2004, this pedigree numbered 18,149 of whom 15,409 were still living.[14]

The blood samples from the pedigree were shipped to James Gusella's laboratory at Massachusetts General Hospital. In only four years, by mid-1983, Gusella had located a region near the end of the short arm of chromosome 4 that was close to the gene,[15] but, given the technological limitations at the time, it took another ten years to find and sequence the gene itself.[16] Some idea of the enormous amount of work that went into locating and characterizing the Huntington's gene is apparent from the authorship of the paper, which is given simply as "The Huntington's Disease Collaborative Research Group." It turns out this is the collective title for six groups located at different institutions. There are multiple named authors from each institution.

The nature of the defect in Huntington's chorea was unexpected. In the middle of the gene is the sequence CAG. The four bases in DNA are cytosine (C), guanine (G), adenine (A), and thymine (T). In the genetic code, they are read in groups of three. Each base is attached via a sugar molecule (deoxyribose) to a phosphorous group that hooks the whole structure into the DNA backbone (see Figure 1–1). This structure is referred to as a nucleotide with a group of three nucleotides being a trinucleotide. The CAG sequence specifies the amino acid glutamine in the middle of a nerve cell protein that was named huntingtin and was encoded by the Huntington's gene. The CAG sequence in the gene and the corresponding glutamine sequence in the protein are repeated a number of times. In Huntington's disease, there are more CAG repeats than normal. The longer the CAG stretch, the earlier the onset of the disease. This type of trinucleotide repeat mutation is not unique to Huntington's disease, but is characteristic of certain other genetic diseases as well (see Table 1–1).

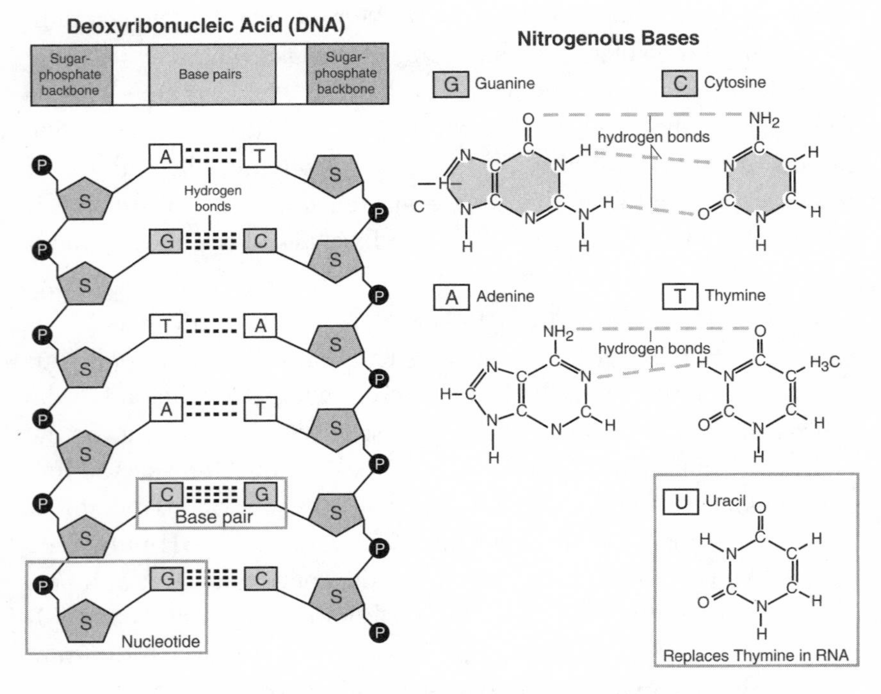

Figure 1–1 Left. A short sequence from the DNA double helix showing the four bases, adenine (A), thymine (T), guanine (G), and cytosine (C). Each base is bonded to a sugar molecule (deoxyribose) which is linked in turn to a phosphate atom to form a nucleotide. The nucleotides are linked to each other via strong, covalent bonds to form the sugar-phosphate backbone of each strand of the helix. The two strands of the helix are held together by hydrogen bonds between the bases with A pairing with T and G with C. Right. G-C and A-T base pairs in more detail. Note that hydrogen bonds are much weaker than covalent bonds and that uracil (U) replaces thymine in RNA.

Courtesy: National Human Genome Research Institute.

Table 1–1 Examples of trinucleotide repeat diseases

		Number of repeats	
Disease	**Trinucleotide**	**Normal**	**Disease**
Huntington's disease	CAG	10–35	40–121
Fragile X syndrome	CGG	5–54	>200–2,000
Friedreich's ataxia	GAA	7–34	200–1,700
Myotonic dystrophy	CTG	5–37	50–11,000

*Huntington's disease is one of eight "polyglutamine diseases" where the sequence CAG is amplified. Each of these sequences occurs in a specific gene encoding a different protein and the disease is neurological in every case. Although the other three diseases shown each involves a specific trinucleotide, these are not within coding regions of the respective genes. Hence, they do not specify different amino acids in the protein products of these genes, although they would do so if they fell within the coding regions. For each disease, there is a numerical gap between the number of repeats found in a normal individual and the number required to cause the disease. This reflects existing uncertainty as to the number of repeats required for the disease to express itself.

Ethnicity, religion, and the gene-hunting companies

As Nancy Wexler's Venezuelan pedigree for Huntington's chorea shows, certain populations are particularly suitable candidates in the search for disease genes. For example, Mormons are a favorable population for the discovery of new disease genes. Mark Skolnick, a University of Utah scientist, realized this many years ago when he became interested in the genetics of breast cancer. In 1991, he was one of the founders of Myriad Genetics of Salt Lake City, Utah, and is currently the chief scientific officer of the company.[17]

Utah's Mormon population was established in the 1840s. Its founders were often polygamous, had large families, and seldom moved. Mormons also marry young so the time span between generations is relatively short. Furthermore, they keep meticulous genealogical records and Myriad Genetics has the rights to these records. As a result, the company has been involved in the identification of the two genes involved in 80% of hereditary breast cancer cases (*BRCA1*, *BRCA2*) as well as genes important in prostate cancer, colon cancer, melanoma, and also genes disposing individuals to risks other than cancer. The company has developed predictive tests for these genes. Their gene-searching technology is especially useful for such cancer genes because, in addition to having available a computerized genealogy of Mormon pioneers and their descendants, the company can access the Utah Tumor Registry. The registry has required that a record be kept of every cancer occurring in the state since 1973. Using these tools in combination has proved a powerful way of finding tumor genes.

The molecular diagnostic products marketed by Myriad Genetics are "designed to analyze genes and their mutations to assess an individual's risk for developing disease later in life or a patient's likelihood of responding to a particular drug, assess a patient's risk of disease progression and disease recurrence, and measure a patient's exposure to drug therapy to ensure optimal dosing and reduced drug toxicity."[18] Myriad's molecular diagnostic revenues in 2009 were $326.5 million, a 47% increase over the previous year. One reason for this profitability is that Myriad is currently the exclusive provider of tests for the *BRCA1* and *2* genes, which are protected by patents. These patents, which are under legal challenge (Chapter 5, "Genes and cancer"), prevent gene-testing companies like 23andme or deCODE genetics from offering tests for these genes, although they do offer tests for other potential genetic risk factors for breast cancer. This is a potential source of confusion for the uninitiated because these companies assess some, but not all, of the potential breast cancer genetic risk. This is why it is of critical importance for people sending off a saliva or cheek swab sample to a gene-testing company to consult with professionals who can inform them as to exactly what the tests will and will not tell them.

One advantage the Mormon population lacks for gene hunters is that, unlike the Amish, it is not inbred.[19] Although Brigham Young founded Salt Lake City in July 1847 with around 2,000 followers, colonization proceeded rapidly so that by 1890, when most immigration into Utah had ended, there was a total population of 205,889, of whom about 70% were Mormons. They included a great many Mormon converts who frequently came from Great Britain and Scandinavia.

Certain populations benefit the gene hunter by originating from small founding populations. Just by chance, this sometimes means that a deleterious gene may be amplified in the founding population as compared with the population from which it was derived. For example, Tay-Sachs is a common genetic disease among Ashkenazi Jews. Suppose you have a settlement containing 100 couples (excluding children for the purpose of the example) giving a population of 200. Suppose that 10 individuals are carriers of the gene. This yields a carrier frequency of 5%. If 10 couples from the original population decide to form a new settlement and, by chance, they include the 10 carriers, the frequency of carriers rises to 50%. This is the founder effect and accounts for the high frequency of some genetic diseases among certain populations (see Chapter 3, "Ethnicity and genetic disease").

When such a population can be identified and has remained relatively homogenous, it becomes an attractive target for a gene-hunting company. At the Genome 2001 Tri-conference held in San Francisco in March 2001, Phillipe Douville, Vice President and Chief Business Officer of Galileo Genomics Inc. of Montreal, remarked that the "main recognized founder populations in the world are those of Quebec, Finland, Sardinia, Iceland, Costa Rica, the northern Netherlands, Newfoundland, and several discrete ethnic groups including the Ashkenazi Jews."[20]

In fact, it was the population of Quebec that Galileo, now called Genizon BioSciences, planned to focus on. About 15,000 French settlers arrived in eastern Canada in the course of the seventeenth century.[21] Around 2,600 of these hardy souls made their way to Quebec. This population has expanded 800 times over the ten generations since, with intermarriage within the group predominating. Thus, the Quebec founder population is relatively homogenous. Genizon BioSciences claims to have over 47,000 subjects in its biobank, 95% of whom have authorized the company to contact them again.[22] Genizon has research teams investigating eight complex conditions, including Alzheimer's disease, obesity, and schizophrenia by means of genome wide association studies. The company hopes to identify specific patterns of genetic variation that correlate with these conditions and, ultimately, to tie each condition to specific genetic markers (see Chapters 3 and 4 for a fuller discussion).

Gene hunting can be an expensive and unprofitable business. It is usually supported for a while by grants, contracts, venture capital, and deep-pocketed drug companies, but, eventually, it must be profitable. The fate of IDgene Pharmaceuticals Ltd., a Jerusalem genomics start-up founded in 1999, shows what can happen when the path to profitability is not achieved soon enough.

The founder effect, coupled with homogeneity and inbreeding, means that the Ashkenazi Jews are favorable material for the discovery of new disease and susceptibility genes. Hence, IDgene Pharmaceuticals' goal was to search for disease genes among the Ashkenazim.[23] Suitable patients with major chronic ailments that had four Ashkenazi grandparents were asked to donate a single blood sample for genetic testing. Written consent was required and the results were kept anonymous. Israeli Ashkenazi Jews suffering from asthma, type 2 diabetes, schizophrenia, Parkinson's disease, Alzheimer's disease, breast cancer, and colon cancer were studied. Using this method, the company's president, Dr. Ariel Darvasi, and colleagues reported strong genetic evidence supporting the hypothesis that a gene called *COMT* encoding an enzyme involved in the breakdown of certain neurotransmitters is involved in schizophrenia (see Chapter 6, "Genes and behavior").[24] But, subsequently, IDgene failed to raise sufficient capital to continue in operation and closed down in 2004.[25]

The biggest pedigree of all: deCODE genetics and the Icelandic population

A company called deCODE genetics initiated the biggest gene-hunting project of all time. The company proposed to use the entire population of Iceland as a genetic resource because Iceland was founded by a small group of Scandinavian settlers centuries ago. The population is homogenous, and has undergone many population constrictions.

Irish monks, the first inhabitants of Iceland, arrived in the eighth century, but did not become established permanently.[26] A small band of Norsemen who settled Iceland between AD 870 and AD 930 followed them. In the latter year, an annual parliament, the Althing, was established to make laws and solve disputes, making the Althing the oldest parliament in the world. In 1000, Iceland adopted Christianity as its official religion. Iceland's rule over the intervening centuries has been complex, beginning with its recognition of the King of Norway as its monarch in 1262–1264 and ending with a complete dissolution of Iceland's ties with Denmark in a 1944 referendum.

Viking traders brought the black plague to Iceland. The disease killed as many as 40,000 inhabitants or more than half the population between 1402 and 1404. The plague returned in 1494–95 with a similarly devastating effect. Around 15,000 people, one-third of the population, died during the smallpox epidemic of 1707–09 just as the Icelandic population was recovering from the depredations of the plague and farming was beginning to flourish. In 1783, the Lakigigar eruption resulted in one of the world's worst volcanic disasters. The eruption lasted for eight months. Gases from the eruption reached altitudes of greater than 9,000 feet. The aerosols formed by these gases cooled the Northern Hemisphere by as much as 1 degree centigrade. The haze that formed caused the loss of most of Iceland's livestock from eating fluorine-contaminated grass. Crop failure from acid rain also occurred resulting in the death of 9,000 people, about one-quarter of the population, from the resulting famine.

The small founding population of Iceland coupled with the population bottlenecks just described, plus the relative isolation of the Icelandic population from immigration, rendered it a natural laboratory for human genetic research. In 1996, Kari Stefansson, a native Icelander and Chief of Neuropathology at Boston's Beth Israel Deaconess Hospital, left his comfortable academic perch to found deCODE genetics, a company whose goal was nothing less than to use the enormous human genetic database of Iceland to identify genetic factors involved in common ailments.[27] His certainty that multiple sclerosis involved such factors and his frustration in trying to identify them was one of the underlying reasons for this move.

Genealogy is a passion in Iceland and local newspaper obituaries give detailed family trees that can extend back a hundred years or more. Furthermore, comprehensive clinical records of Iceland's public health service go back as far as 1915. Stefansson recognized that a computerized database of this information for the entire Icelandic population would be an invaluable tool for tracking down genetic diseases. Even more important, Stefansson knew that an exclusive agreement between his company and the government of Iceland would be an integral part of any business plan. This would give deCODE a major advantage over potential competitors.

In February 1998, deCODE signed an agreement with Hoffman-La Roche stipulating that Hoffman-La Roche would pay deCODE more than $200 million in "benchmark" payments over five years if the company succeeded in identifying genes associated with common debilitating and often lethal syndromes like stroke, heart disease, Alzheimer's disease, and emphysema.[28] However, these "benchmark" payments required that deCODE achieve specific goals within a given amount of time. In an ominous portent of things to come, deCODE failed to achieve the expected goals and received only around $74.3 million of the original total.

The company initially began its work with DNA donated by small groups of Icelanders.[29] This approach was followed up by a publicity campaign designed to attract donors in larger numbers. But the great coup was the Althing's passage of the Health Sector Database Act in December 1998 by a majority of 37 to 20 with 6 abstentions and with the strong support of the Prime Minister David Oddsson.[30] The database act authorized the development of a Health Sector Database for the collection of genetic and medical information already stored in various places around Iceland as part of the country's national health system.

The government had several altruistic reasons for wanting to form the database.[31] First, the act stated that the comprehensive medical records held by the national health system were a national resource that should be kept intact and utilized in the best way possible. Because government funds were used to support construction of the database, the government rejected the notion that any records submitted to the database could be of a proprietary nature. Neither legal entities nor individuals could be granted ownership of specific medical data. Hence, the database would provide the nation with the opportunity to make use of its information to improve medical services for the people of Iceland.

Second, in 1997, the Ministry of Health and Social Security made public a policy statement regarding its plans for utilizing information technology within the national health system. The idea was to create a number of dispersed personal databases that could be linked. This linked database would include medical records and summarize research in fields of possible relevance to Icelandic health, including epidemics, demographics, and genetic diseases. The cost of constructing such a database was beyond the capacity of the national government, but deCODE's participation would make the effort possible.

Third, the government hoped the database might reverse the Icelandic brain drain by enticing Icelandic scientists interested in human genetics to return to their country. Fourth, the government expected that the database would provide economic benefits to Iceland.

Further actions favorable to deCODE genetics followed.[32] In January 2000, the minister of health granted a 12-year license to the company to operate the database. In 2002, the Althing passed a bill permitting the government to issue state bonds as security for a $200 million loan to deCODE to show its support for the company and to help in financing construction of the database.

Initially, the idea of establishing such a database met with strong support as the results obtained held the potential of bringing to Iceland enormous sums of money from pharmaceutical companies. Several Icelandic politicians expressed the hope that the deCODE database might be as significant for the country as the discovery of North Sea oil was for Norway.[33] Opposition to the project soon emerged, however, as it became evident that Iceland would be the only country in the world to have passed a law authorizing a private company to collect, store, and analyze the genetic heritage of an entire population for commercial purposes.

Some of the concerns were as follows: First, if an individual's personal health information was accessed from the database by an unauthorized person or company, that individual's privacy would be violated or worse.[34] deCODE countered that a person's information would be encrypted. Second, the database act assumed all Icelanders had given their consent to have their personal statistics entered. Although an individual could opt out of the database at any time, data already recorded on that person remained in the database. Furthermore, Icelanders had only six months from the time that the database was constructed to request that their data not be included in the database. This provision was only added to the act because an earlier version had assumed "presumed consent" rather than informed consent. Additionally, data relating to deceased family members would be included automatically without regard to the possible privacy interests of living relatives.

Third, there was danger of genetic stereotyping. One of the diseases studied for which Hoffman-La Roche provided financing was schizophrenia. If a certain fraction of the population proved to have or be susceptible to this disease, then this might suggest to health insurers that anybody of Icelandic heritage any place on earth might be at risk of becoming schizophrenic. Fourth, as the sole licensee, deCODE had monopoly control of the data, although the database itself was the property of the national health system and was managed by the government. Furthermore, deCODE was to be permitted to use the data for commercial purposes for 12 years and access to the data by others was denied if it threatened the financial interest of the company. Fifth, deCODE would make its data available to pharmaceutical and insurance companies for a price. Furthermore, the arrangement with Hoffman-La Roche, according to which deCODE would exclusively investigate 12 different diseases, prevented others from studying these diseases in Iceland.

Pétur Hauksson, a psychiatrist, founded Mannvernd (an Icelandic word meaning human protection), a nonprofit human rights group. Its goal soon became to overturn the Health Sector Database Act. One of Mannvernd's most important complaints was that the act was based on the presumed consent of Icelandic citizens. In addition, citizens who agreed to give blood for one of deCODE's genetic disease investigations had to consent to have the samples used for other genetic studies without knowledge of what they might be. Because of Mannvernd's efforts, Icelanders were now able to refuse to have their information entered in the database by submitting an appropriate form. By June 2001, 20,000 Icelanders, about 7% of the population, had opted out of the Health Sector Database. The Icelandic Medical Association also voiced its opposition to the database act. Many doctors refused to turn over patients' records without their consent. In April 1999, the Icelandic Medical Association brought the Health Sector Database Act before the World Medical Association. The latter body stated full support for the position taken by its Icelandic member in opposition to the database act. Other international criticism was also on the rise. For example, Harvard's Richard Lewontin, a distinguished population geneticist, published an op-ed piece in the *New York Times* on January 23, 1999, titled "People Are Not Commodities," which argued that the database act had transformed the "entire population of Iceland into a captive biomedical community."[35]

A major concern of the Icelandic Medical Association was the protection of personal data under the database act. Were the encryption technologies sufficient to prevent some unauthorized individual from linking medical data with a specific individual? The association hired Ross Anderson, a Lecturer in the University of Cambridge Computer Laboratory, in fall of 1998 to look into this question. Anderson concluded that deCODE and the Icelandic Data Protection Commission would have to use coded identifiers that would permit linkage of personal data to specific individuals. Because the encryption system would be broken sooner or later, it seemed to Anderson that informed consent standards would have to apply.

Meanwhile, deCODE had begun to achieve scientific success with the more traditional approach by making use of family pedigrees with their informed consent. Hence, the company's obligations under the database act became more of a burden than an opportunity, especially because deCODE was unable to bring the Icelandic Medical Association and the Data Protection Commission on board. The final blow to construction of the database came on November 27, 2003, the day that the Icelandic Supreme Court rendered its verdict in the case of *Gudmundsdóttir v. Iceland*.

The case was prompted by a young woman who wrote to the Icelandic Ministry of Health in February 2000 requesting that any information in her father's medical records and any genealogical or genetic data concerning him not be transferred to the database. The medical director of health denied her request after he had obtained a legal opinion. The Icelandic District Court upheld the director's decision arguing that the medical information available in the database could not be connected to a specific person. But the Supreme Court reversed the lower court decision stating that Gudmundsdóttir had a personal privacy interest in her father's medical data. However, the Court broadened its ruling pointing out that, because by Icelandic law individual medical records were required to contain detailed information on people's health, employment, lifestyles, social circumstances, and so on, a guarantee had to be applied to ensure the individual's freedom from interference with privacy, home, and family life.

Although the database act was dead, deCODE was making good scientific progress in gene discovery. On its Web site, the company claimed to have "discovered risk factors for dozens of common diseases ranging from cardiovascular disease to cancer."[36] deCODE also introduced a new program called deCODEme, which offered customers complete scans that would allow them to discover their "genetic risk for 46 diseases and traits ranging from heart attack and diabetes to alcohol flush reaction and testicular cancer."[37] The company also offered a cardiovascular risk scan, a similar scan for seven common cancers, and a scan of a person's DNA to discover their genetic roots. The problem was that deCODE had never made a profit, was losing money, and was becoming increasingly indebted to its creditors. On November 17, 2009, deCODE filed for bankruptcy under Chapter 11 of the United States Bankruptcy Code.[38] At the same time, it entered into an agreement with Saga Investments LLC to purchase its Iceland-based subsidiary Islensk Erfdagreining and its drug discovery and development programs. Following the sale of these assets, deCODE genetics would be liquidated.

In reporting the bankruptcy of deCODE genetics, the *Times* of London said that it had been assured by Kari Stefansson "that ownership of genetic data remained with the company's customers and that Saga would be bound by a privacy policy that prevents disclosure of data to third parties such as insurers, employers or doctors."[39] But Dan Vorhaus, a lawyer with the American firm of Robinson, Bradshaw, and Hinson, which specializes in genomics, was not convinced. He noted the agreements that deCODE had made with its customers were "often unclear and contradictory."[40]

"The ownership is going to change, and the people making decisions about how to run the company are going to change," Vorhaus said. "This information was held by deCODE, a scientific research organisation. What you have now is Saga, an investment company with a different agenda, very much focused on the bottom line.

Within the range of allowable uses, deCODE's new ownership may choose to use that information in a different way, and possibly to a greater extent, than was previously the case."[41]

So the question of genetic privacy, that became such an issue after the passage of the database act, arises once more with the bankruptcy of deCODE genetics. It will become an issue again should other gene-hunting companies declare bankruptcy or enter into mergers or takeovers such as the one between deCODE and Saga. In January 2010, deCODE emerged from bankruptcy under the ownership of Saga Investments.[42] Its new CEO was a lawyer named Earl Collier with its founder and former CEO Kari Stefansson now head of research.

How many disease genes are there?

In 1957, Victor McKusick was appointed director of the new Moore Clinic for Chronic Diseases at Johns Hopkins University and head of the newly established Division of Medical Genetics at its medical school.[43] He had come into human genetics via his research on disorders affecting connective tissue, including Marfan's syndrome. Marfan's sufferers typically have long slender limbs and are often taller than normal. The most serious conditions associated with the disease primarily involve the cardiovascular system, as there may be leakage through the mitral or aortic valves that control blood flow through the heart. McKusick noticed that Marfan's syndrome exhibited a familial pattern of occurrence and, indeed, we know today that a dominant genetic mutation is involved. The Marfan's pedigree sparked McKusick's interest and he began to specialize in human clinical genetics.

In 1966, he published his first catalog of all known genes and genetic disorders, *Mendelian Inheritance in Man* (MIM). The 12th edition of his catalog was published in 1998. Meanwhile, a free online version (OMIM) first became available in 1987. It is continuously updated. The database is linked with the National Center for Biotechnology Information and the National Library of Medicine for distribution. In the 1980s, only a few genes were being found each year. By 2000, the number of genes discovered each year was approaching 175. More than 6,000 single gene disorders are currently known,[44] meaning that mutations in somewhere around 24% of the approximately 25,000 human genes found so far can cause genetic disease. Because of the broad interest in disease genes as well as the availability of increasingly sophisticated technical and statistical tools, the rate of disease gene discovery has expanded rapidly. Whether or not it plateaus at some point remains to be seen.

It was originally thought that the human genome might contain as many as 100,000 genes. Once the Human Genome Project was completed in 2003 and a few further revisions were made, this number dropped to around 25,000, roughly the same range as the mouse (see Table 1–2). But the surprising thing is that these protein-encoding genes represent less than 2% of the 3.2 billion base pairs in the human genome.[45] Unlike the even spacing of a string of pearls, our genes often cluster in gene-rich regions separated by gene-poor deserts.

Table 1–2 Genome sizes and gene density in humans as compared with other organisms frequently used in genetic research

Organism	Estimated size (base pairs)	Estimated gene number	Average gene density	Chromosome number
Homo sapiens (human)	3.2 billion	~25,000	1 gene per 100,000 bases	46
Mus musculus (mouse)	2.6 billion	~25,000	1 gene per 100,000 bases	40
Drosophila melanogaster (fruit fly)	137 million	13,000	1 gene per 9,000 bases	8
Arabidopsis thaliana (plant)	100 million	25,000	1 gene per 4,000 bases	10
Caenorhabditis elegans (roundworm)	97 million	19,000	1 gene per 5,000 bases	12
Saccharomyces cerevisiae (yeast)	12.1 million	6,000	1 gene per 2,000 bases	32
Escherichia coli (bacteria)	4.6 million	3,200	1 gene per 1,400 bases	1
H. influenzae (bacteria)	1.8 million	1,700	1 gene per 1,000 bases	1

From Human Genome Project Information: Functional and Comparative Genomics Fact Sheet. www.ornl.gov/sci/techresources/Human_Genome/faq/compgen.shtml

The human genome is distributed between 23 chromosomes. These are found singly in sperm and eggs (haploid), but in pairs in all of the rest of our cells (diploid). This halving in chromosomes number in eggs and sperm is achieved during the two cell divisions of meiosis. During the first division, homologous paternal and maternal chromosomes pair respectively with paternal and maternal chromosomes assorting independently of each other. During the pairing, chromosome segments are exchanged between homologs, a process called genetic recombination (see Glossary for a brief introduction to Mendelian genetics). Although not generating new genetic alterations, the processes of independent assortment and recombination provide the opportunity to assort existing parental genes in a variety of new combinations. Creation of all of this new genetic variability on which natural selection can act is a major reason why sexual reproduction predominates in animals and plants.

Like the genes of other higher organisms, human genes themselves are not single blocks of DNA that encode specific proteins. Instead, they are broken up into coding sequences (exons) and noncoding sequences (introns). Following the process of transcription, when the information in a gene is copied into a messenger RNA molecule, the intron sequences are spliced out of the message so only the coding sequences in the messenger RNA can be translated into protein sequence.

What is all that other DNA doing that has no obvious genetic function? We know that at least 50% of the genome, perhaps more, is made up of repeated sequences that do not encode human proteins and often no proteins at all. These repeats are of several kinds, but the most abundant are "mobile" genetic elements that make up roughly 43% of the mammalian genome.[46] They either are or at one time were capable of movement from one site in the genome to another.

Transposons are the first group of mobile elements. They comprise around 3% of the genome. The name transposon evokes the word transposition and, indeed, these elements are capable of moving from one to another place in the genome. The easiest way to think about transposition is as a "cut-and-paste" process. One cuts out a word, or a group of words, in a text and then pastes those words into a specific place elsewhere in the text. The important difference between transposition and cutting and pasting is that, although transposition will take place only into its target DNA sequence, the element can be pasted into that sequence anywhere in the genome. An enzyme called a transposase encoded by the element catalyzes the transposition process. Hence, transposons are sometimes called jumping genes.

The second group includes several sets of elements of which three are the most abundant. The first are endogenous retroviruses. These are viruses whose genetic material is RNA. An enzyme called a reverse transcriptase encoded by the virus catalyzes synthesis of DNA copies of the viral RNA. These DNA copies are then inserted into the genome. The AIDS virus is the best-known retrovirus, but unlike AIDS, the retroviral fragments that inhabit our genomes today are, for the most part, the remains of ancient retroviruses that have lost their ability to become independent of the genome.

LINES (long interspersed nuclear elements) comprise the second group (see Glossary for a more complete discussion of LINES and SINES). They are retrotransposons. One way to think about a retrotransposon is as an odd sort of printing press. An RNA copy is transcribed from the retrotransposon DNA. In the case of LINES, translation of the RNA copy results in the production of two proteins. One of these proteins is essential for the transposition process. The second catalyzes synthesis of a DNA copy of the RNA and then makes a cut in a specific DNA sequence (e.g., TTTTAA/AAAATT for L1) in the genome where the newly made retrotransposon can insert. This method of reproduction has the potential for enormously amplifying the number of retrotransposons in the genome that can then home into their target sequences wherever they are in the genome.

There are several different kinds of LINE elements, but L1, which predominates in the human genome, has evolved along with the mammals over the past 160 million years or more. Expansion in the number of L1s in the genome was rapid, but appears to have slowed down about 25 million years ago. The 500,000 or so copies of L1 present today in the human genome amount to around 18% of its content. The intact L1 element is about 6,000 base pairs in length, but truncated versions are common. L1s are the only active transposons in the human genome today.

SINES (short interspersed nuclear DNA elements) are short DNA sequences of less than 500 base pairs. SINES do not encode any proteins and are not autonomous. They can only transpose with the aid of the two proteins made by active LINE elements. The most important SINES are the Alu elements.[47] More than a million copies of these short DNA sequences are found in the human genome. They represent around 13% of the total DNA. Alu elements originated and coordinated their amplification with the radiation of the primates about 65 million years ago.

Because nobody is exactly sure why human and other animal and plant genomes contain so many repeated elements, they have sometimes been treated as irritants with regard to the real genes, gaining them epithets such as "junk DNA" and "selfish DNA." In a recent review, Goodier and Kazazian point out that a more sophisticated name "dark matter" is coming into vogue for these repeated elements, acknowledging the fact that we don't really understand whether they have an as-yet-to-be-discovered function. Goodier and Kazazian prefer to think of mobile elements as "dark energy."[48] They are "a dynamic force that not only accelerates expansion but also helps set the warp and weft of genomes for better and for worse. Transposable elements arose as intracellular parasites that became domesticated."

Well not entirely. Transposition of these elements can disrupt gene function. In a 1998 paper in *Nature,* Kazazian and his colleagues reported two unrelated cases of hemophilia A for which there was no family history, suggesting that the mutations had arisen de novo.[49] Each of them involved the insertion of L1 sequences into the *F8* gene. So we end this chapter where we began it—with hemophilia. Transposon insertions have also been implicated in a wide spectrum of genetic diseases other than hemophilia.[50]

2

How genetic diseases arise

In 1902, Archibald Garrod (later Sir Archibald Garrod), a casualty physician at St. Bartholomew's Hospital in London, published a paper of remarkable importance. It was titled "The Incidence of Alkaptoneuria: A Study in Chemical Individuality."[1] Garrod had long been interested in compounds that were excreted in the urine and especially in diseases that caused it to change color. Garrod had also struck up a friendship with the distinguished physiological chemist Frederick Gowland Hopkins at nearby Guy's Hospital. They collaborated in the 1890s on several papers describing compounds present in urine.

Garrod first encountered patients with alkaptoneuria in 1897. The urine produced by persons having this disease turns black upon exposure to air. Garrod knew the disease was not serious. However, he believed that alkaptoneuria might be inborn because individuals had the disease for life. The compound that turned black in alkaptoneurics' urine had been shown earlier to be homogentisic acid.

If alkaptoneuria was inherited, one might expect to see the disease in more than one family member. In a 1901 article, "About Alkaptoneuria," Garrod reported the key observation. The mother of one of his alkaptoneuric patients had her fifth child, a baby boy, on May 1, 1901. Garrod instructed the attending nurses to examine the baby's diapers carefully for the telltale staining that would indicate the presence of homogentisic acid in the baby's urine. No staining was evident 15 hours after the boy's birth, but by 52 hours after the infant's birth, deep staining was evident. Thus, alkaptoneuria appeared to be an inherited disorder. Garrod's finding that first-cousin marriages tended to yield higher frequencies of alkaptoneurics also supported this conclusion.

Garrod did not know about Mendel's paper. It had gathered dust since it was published in 1866, but was rediscovered independently in 1900 by Hugo de Vries in Holland and Karl Correns in Germany.[2] However, William Bateson, a fellow at St. John's College, Cambridge, had learned about Mendel's paper from de Vries. Bateson had been studying discontinuously varying characters (e.g., white versus black coat color in mammals) for years and had laid out his findings in detail in his book *Materials for the Study of Evolution* (1894). But until Mendel's paper came along, Bateson had no way of relating his findings to a model of inheritance. Mendel's paper provided that model and Bateson became a major defender of Mendel. Defender is the right word because in 1897, prior to the rediscovery of Mendel's principles, Francis Galton had proposed an alternative model of inheritance.

Galton's Ancestral Law of Heredity seemed to work well for continuously varying characters such as height because it posited, reasonably enough, that each parent contributes one-half of their hereditary material to each child, each grandparent contributes one-quarter, and so forth.[3] But it was hard to see how it could be made to work for discrete characters like the ones studied first by Mendel and later by Bateson. A great war of words ensued in the scientific periodicals between Bateson, Mendel's champion in Britain, and Galton's disciples Karl Pearson, a superb statistician, and Walter Weldon, a fine zoologist. Mendelism eventually triumphed, but it got so bad that Bateson behaved very rudely on July 1, 1908, at the Darwin-Wallace celebration of the Linnean Society of London.[4] Some "wag on the Linnean" executive sat Pearson and Bateson next to each other. Pearson planned to greet Bateson politely, if not warmly. But Bateson sat down in his chair sideways and remained with his back to Pearson during the whole of the ceremony.

A few months after Bateson had learned about Mendel's paper, he became aware of Garrod's work. Bateson realized that alkaptoneuria might be caused by a recessive gene mutation, and, in a paper presented to the Evolution Committee of the Royal Society on December 17, 1901, he elaborated on this hypothesis.[5] Bateson and Garrod now entered into a correspondence that led to Garrod's 1902 paper on alkaptoneuria and its probable inheritance. Garrod went further by pointing out that albinism and cystineuria might also be genetic diseases.

In 1908, Garrod delivered the Croonian Lectures before the Royal College of Physicians.[6] He titled his topic "inborn errors of metabolism." The lectures were published as a book of the same title a year later. A second edition appeared in 1923. In the later edition, Garrod explained that he thought alkaptoneurics accumulated homogentisic acid because of an enzymatic defect that prevented them from breaking the substance down. As we now know, Garrod was correct. Because of an enzymatic defect, alkaptoneuria blocks a step in the pathway that leads to the degradation of the amino acid phenylalanine. Other genetic diseases interfere with different steps in the pathway. The best known of these diseases is phenylketoneuria. This genetic disease causes mental retardation unless the phenylketoneuric child is reared on a diet low in the amino acid phenylalanine.

The significance of Garrod's work was only slowly recognized by geneticists. They were mostly engaged in deciphering the laws of heredity with the aid of model systems, most notably fruit flies and maize. But these systems were not very suitable for studying the relationship between genes and biochemical pathways so George Beadle and Edward Tatum used a different model system, the red bread mold, *Neurospora crassa*. Its simple requirements allowed them to isolate mutations blocking biochemical pathways to the formation of compounds like amino acids and vitamins. Thus, it became apparent to them that each gene probably specified a single protein. This led to the "one gene one enzyme" hypothesis, an expression coined by Norman Horowitz, also working with *Neurospora*. We now know that some enzymes and other proteins are composed of polypeptide chains encoded by different genes. Hemoglobin is a good example. Adult hemoglobin (there are other kinds) is composed of two alpha and two beta polypeptide chains encoded by different genes. So the hypothesis has now been condensed to the one gene-one polypeptide hypothesis.

In 1958, Beadle and Tatum were awarded the Nobel Prize in Physiology or Medicine for their discovery that genes specify proteins, but Beadle was characteristically modest in his acceptance speech.

"Regardless of when it was first written down on paper, or in what form, I myself am convinced that the one gene-one enzyme concept was the product of gradual evolution beginning with Garrod and contributed to by many others including Moore, Goldschmidt,

Troland, Haldane, Wright, Grüneberg, and many others. Horowitz and his co-workers have given it...its clearest and most explicit formulation."[7]

In 1996, a group of Spanish investigators reported the isolation of the gene defects in which cause alkaptoneuria. The *HGD* gene (also called *HGO*) encodes the enzyme (homogentisate 1,2 dioxygenase). This enzyme converts homogentisic acid to maleylacetoacetic acid, the next compound in the degradation pathway of phenylalanine.[8] They also characterized some of the mutations blocking the function of this enzyme. Thus, Garrod's original hypothesis, first published in 1902, that alkaptoneuria was the result of some kind of genetic defect, was finally proved unequivocally only in 1996.

Having established that genes encode proteins and that mutations affecting these proteins can cause genetic diseases, the rest of this chapter is devoted to the subject of how genetic diseases arise. As we shall see, many of these diseases result from mutation, whereas others arise from imbalances in chromosome number and other alterations.

Most genetic diseases are caused by mutations

So what were those mutations in the *HGD* gene? Most of them were missense mutations.[9] What are they? To explain missense and other mutations, a short detour into the properties of the genetic code is in order.

As soon as it became apparent that DNA was the genetic material, the question became how the four bases in DNA—adenine (A), thymine (T), guanine (G), and cytosine (C)—could specify the 20 different amino acids commonly found in proteins. Obviously, they had to be used in combination. But all combinations of two yield only 16 possibilities, too few, while all combinations of three yield 64 different arrangements, too many. It turned out that the latter solution was correct because the genetic code is degenerate (see Table 2–1). That is, all the amino acids except for two are specified by more than one sequence of three bases (codon). A missense mutation results from a base pair substitution in the DNA that causes a codon change such that the original amino acid in the protein coded by the gene is replaced by a different amino acid. This often causes the protein to become nonfunctional or poorly functional. Thus, one of the missense mutations in the *HGD* gene causes an AUG codon specifying methionine to change to a GUG codon that will be read as valine (see Table 2–1). Note that uracil (U) replaces thymine (T) in RNA, but the two bases have very similar properties. Hence, codons are normally given in terms of their RNA bases. That is, GUG rather than GTG.

Table 2–1 The genetic code showing the RNA codons

Second Position

First Position	U		C		A		G		Third Position
U	UUU	Phe	UCU	Ser	UAU	Tyr	UGU	Cys	U
	UUC		UCC		UAC		UGC		C
	UUA	Leu	UCA		UAA	*Stop*	UGA	*Stop*	A
	UUG		UCG		UAG	*Stop*	UGG	Trp	G
C	CUU	Leu	CCU	Pro	CAU	His	CGU	Arg	U
	CUC		CCC		CAC		CGC		C
	CUA		CCA		CAA	Gln	CGA		A
	CUG		CCG		CAG		CGG		G
A	AUU	Ile	ACU	Thr	AAU	Asn	AGU	Ser	U
	AUC		ACC		AAC		AGC		C
	AUA		ACA		AAA	Lys	AGA	Arg	A
	AUG	Met	ACG		AAG		AGG		G
G	GUU	Val	GCU	Ala	GAU	Asp	GGU	Gly	U
	GUC		GCC		GAC		GGC		C
	GUA		GCA		GAA	Glu	GGA		A
	GUG		GCG		GAG		GGG		G

Amino acid abbreviations: Ala = alanine; Arg = arginine; Asn = asparagine; Asp = aspartic acid; Cys = cysteine; Gln = glutamine; Glu = glutamic acid; Gly = glycine; His = histidine; Ile = isoleucine; Leu = leucine; Lys = lysine; Met = methionine; Phe = phenylalanine; Pro = proline; Ser = serine; Thr = threonine; Trp – trlyptophan Tyr = tyrosine; Val = valine

Source: http://tigger.uic.edu/classes/phys/phys461/phys450/ANJUM02/codon_table.jpg

But how is the genetic code punctuated? It must have some sort of punctuation because the protein synthesizing machinery has to know where to begin to translate the information contained in the messenger RNA into its protein product and when to stop. Maybe the code is punctuated like an English sentence with external punctuation: the capital letter at the beginning of the sentence, a period at the end, and internal punctuation as well, commas, semicolons, and so forth. In fact, the genetic code has no internal punctuation. A sentence in the genetic code begins with AUG, which specifies the amino acid methionine, and ends with one of three codons (UAA, UGA, UAG). They are called stop or termination codons. These are the periods at the ends of genetic sentences.

The existence of stop codons, and the lack of internal punctuation in the genetic code, opens the door for two other kinds of mutations to wreak havoc with the protein product of the gene. The simplest of these mutations to understand is the nonsense mutation. Suppose, somewhere in the middle of the mRNA sequence encoding a protein, there is a CAG sequence encoding the amino acid glutamine. If this sequence is mutated to UAG, the sentence ends prematurely at this genetic period, so synthesis of the protein chain terminates at that point, resulting in an incomplete protein. Chain-termination mutant is a useful descriptive name, but to show the reader how weird geneticists can sometimes be, it is interesting to note in passing that UAG, UAA, and UGA mutants are called amber, ochre, and opal mutations. Why? A bacterial virus (bacteriophage) T4 that infects *E. coli* used to be a favorite genetic tool. A couple of young scientists, Dick Epstein and Charley Steinberg, at California Institute of Technology, who were studying T4 genetics, convinced a graduate student at that institution, Harris Bernstein, to look for a specific kind of T4 mutant they were interested in.[10] Bernstein's reward was to have the mutant named after him. Bernstein found the mutant. Because Bernstein would be a rather awkward name for a mutant, the three decided to call it an amber mutation, the English translation for Bernstein. Bernstein then acquired the nickname "Immer Wieder Bernstein" (i.e., forever amber). The ochre and opal mutations were named to conform to the nonexistent color code established for the amber mutation.

Missense and nonsense mutations are also known as point mutations because they exchange one nucleotide pair (e.g., A-T) for another (e.g., G-C). Silent mutations and neutral mutations are the other two kinds of point mutations. They have no measurable deleterious effects. Silent mutations do not result in an amino acid change. For example, a change from AUU to AUC causes no amino acid change because both codons specify isoleucine. A neutral mutation does result in an amino acid change, but one that does not affect the function of the protein encoded by the gene. The new amino acid will usually have chemical properties that are similar to those of the amino acid specified by the original amino acid.

The absence of internal punctuation in the genetic code means that the insertion or deletion of a base pair within the coding sequence of a gene can cause trouble. In the example given in Table 2–2, the addition of a G-C base pair in the DNA adds a G to the second codon in the mRNA, which is now read as GCA rather than CAU. As a result, the histidine in the second position is changed to alanine and the proline and arginine residues in the third and fourth positions in the normal protein are replaced by two serine residues followed by a stop codon, where synthesis of the protein chain will now terminate. Deletion of a G-C base pair in the third codon causes a C to be deleted from the mRNA so the proline, arginine, arginine, glycine, alanine sequence in the normal protein is now replaced with leucine, valine, glycine, and glutamic acid. Deletion and addition mutations not only involve single base pairs, but often involve lots of them, sometimes extending to whole chromosome regions.

Table 2–2 Consequences of frameshift mutations*

Normal amino acid sequence in protein	His	His	Pro	Arg	Arg	Gly	Ala
Normal codon sequence in mRNA	CAU	CAU	CCU	CGU	AGG	GGA	GCC
Amino acid sequence in protein	His	Ala	Ser	Ser			
Add one base to mRNA sequence	CAU	GCA (+) G	UCC	UCG	<u>UAG</u>	GGG	AGC
Amino acid sequence in protein	His	His	Leu	Val	Gly	Glu	
Delete one base from mRNA sequence	CAU	CAU	CUC (-)C	GUA	GGG	GAG	CC

*Note that the base changes in mRNA are, of course, the result of a base pair addition or deletion in the corresponding DNA sequence. The abbreviations for the amino acids are as follows: alanine (Ala), arginine (Arg), glutaminc acid (Glu), glycine (Gly), histidine (His), leucine (Leu), proline (Pro), serine (Ser), and valine (Val).

The previously described mutations are all what are referred to as single nucleotide polymorphisms or SNPs. The point mutations exchange one base pair for another. The frameshift mutations or indels insert or delete a base pair. Recall that only about 2% of the human genome is actually composed of protein-coding genes. There are literally millions of SNPs in our genomes that have no effect on our genes. However, SNPs may account for over 80% of the genetic variation between individuals.[11]

SNPs, the HapMap, and genome-wide association studies

The implication of the discovery of SNPs for genetic medicine can hardly be overstated. They, together with rapid and cheap genomic sequencing, are allowing scientists and doctors to march toward the era of personalized medicine (see Chapter 10, "The dawn of personalized medicine"). Personalized medicine will permit people to obtain profiles for a vast number of gene variants that increase the risk that an individual will be subject to genetic diseases, especially those that have a complex basis like heart disease or schizophrenia (see Chapters 4 and 6, "Susceptibility genes and risk factors" and "Genes and behavior," respectively). These complex diseases often involve several or more genetic risk factors whose ability to cause disease is frequently subject to a very large environmental component.

However, SNPs by themselves are not of great use without a means of classifying them. This is the goal of the International HapMap Project. It is classifying the millions of existing SNPs into groups that can be used to map risk-factor genes rapidly. HAP derives from the term haplotype. Haplotype is a contraction of haploid genotype. As mentioned in Chapter 1, "Hunting for disease genes," sperm and eggs have one set of chromosomes each and are called haploid. Somatic cells have two sets of chromosomes and are termed diploid. Haplotype refers to a set of genetic markers and molecular markers such as SNPS that are on the same chromosome and usually segregate together.[12]

On July 18–19, 2001, the National Institutes of Health (NIH) sponsored a conference whose title was "Developing a Haplotype Map of the Human Genome for Finding Genes Related to Health and

Disease."[13] By that time, about 2.4 million SNPs had been identified. It was estimated that a comparison of the DNA from two people chosen at random would show that they differed at about 1 in every 1,000 DNA sites. Hence, when these two genomes were compared over their entire extent, there should be about 3 million differences. Eventually, the total number of SNPs will probably total between 10 and 30 million. Of these, perhaps 4 million are relatively common with frequencies over 20%.

In theory, there could be many different haplotypes in a chromosome region, but so far, only a few common haplotypes have been found. To find genes that increase the risk of a particular disease, the SNP profiles of people with and without the disease are compared to see whether one of the profiles correlates with the disease. Occasionally, one of the SNPs will mark the gene itself, but more frequently, they do not. Because the chromosomal locations of many SNPs are known, this makes it possible to map the genetic risk factor to a specific chromosomal location. Once the variant risk-factor SNP or the risk-factor gene is identified, determining whether a person has it or not is very easy.

The International HapMap Project was officially launched at a meeting held from October 27–29, 2002, with a budget of $120 million.[14,15] Funding and workload are divided among the different participating groups of scientists in several countries. For example, in Japan, scientists at the University of Tokyo and the University of Hokkaido collaborate. The first group's assignment is to map SNPs in about 25% of the human genome and, specifically, chromosomes 5, 11, 14, 15, 16, 17, and 19. The second group at Hokkaido provides samples and consultation with the public. The Japanese MEXT (Ministry of Education, Culture, Sports, Science, and Technology) funded this part of the project. In the United States, the NIH is providing funding for six participating groups. Between them, they have divided up chromosomes constituting around 31% of the genome. Assignments were made to participating groups in other countries in a similar fashion.

The participants were to examine a total of 270 samples. Some came from a U.S. Utah population with Northern and Western European ancestry, whereas other samples came from the Yoruba people of Northern Nigeria, Japanese from Tokyo, and Han Chinese from Beijing. Although this might seem to imply that the consortium expected differences that were correlated with race, it makes great scientific sense because these populations have been separated from one another for millennia. Hence, examining them should give a wide range of SNP markers, some of which ought to be more frequent in one population than another.

The consortium published a second-generation HapMap in 2007.[16] Over 3.1 million SNPs had now been mapped. The authors pointed out that the Phase II HapMap "provided the most complete available resource for selecting tag SNPs genome-wide." Despite the success of the HapMap Project, the consortium also acknowledges that genome sequencing is getting ever faster and cheaper. "Finally, in the future, whole-genome sequencing will provide a natural convergence of technologies to type both SNP and structural variation. Nevertheless, until that point, and even after, the HapMap Project data will provide an invaluable resource for understanding the structure of human genetic variation and its link to phenotype."

The HapMap together with commercially available DNA arrays on microchips or beads has made genome-wide association studies (GWAS) easier to conduct. This has greatly simplified attempts to define the genetic basis of complex diseases. A review published in 2008 listed over 50 disease susceptibility loci. For diseases including types 1 and 2 diabetes, inflammatory bowel disease, prostate cancer, and breast cancer, "there has been a rapid expansion in the numbers of loci implied in predisposition."[17]

As a preview of the personalized genetics of the future, let us consider J. Craig Venter's genome. In 1998, Venter, an exuberant biologist and entrepreneur, founded Celera Genomics, whose Latin root means speed. Its goal was to sequence an entire human genome in three years.[18] He was in competition with a government-funded consortium that had split the task up among several different academic centers. James Watson led the consortium first, but was later supplanted by Francis Collins, discoverer of the cystic fibrosis gene. The consortium had begun the project in 1990, but Venter had developed a much faster method for genome sequencing. Venter's group had already completed a number of genomic sequences, principally of the small genomes of bacteria. Thus, Venter and the government were in a race—with Venter, the late starter, progressing rapidly. By June 2000, the competitors were neck and neck preparing draft sequences of the human genome. The consortium and Venter published draft versions of the complete human DNA sequence in 2001. But in 2002, Celera abruptly fired Venter, so the consortium won when, in 2003, it announced that it had completed the first complete human DNA sequence.

By 2006, the irrepressible Venter was up and running again creating the J. Craig Venter Institute (JCVI) by fusing several of the companies he had created earlier. They included The Institute for Genomic Research (TIGR) and The Center for the Advancement of Genomics (TCAG). The alert reader will notice that the acronym TIGR suggests ambition, whereas acronym TCAG corresponds to the four letters in the genetic code.

In 2007, Venter and his colleagues published Venter's own DNA sequence.[19] Some of Venter's DNA sequences were associated with increased risk of antisocial behavior, Alzheimer's, cardiovascular disease, and, somewhat bizarrely, wet earwax. At the time, Huntington Willard, Director of Duke University's Institute of Genomic Research and Policy, called Venter's sequence "the gold standard right now."[20] This was partly because Venter's sequence is the most complete to date, and partly because it was obtained from a single individual rather than from a mixture of donors as was true in the case of the consortium. The public availability of Venter's genomic sequence might not have bothered him, but it could have concerned another person. Some might like to know whether they are at risk for heart disease or Alzheimer's, especially if preventative treatments are available. But would you really want to know if you have antisocial tendencies and do you care if you have wet earwax?

What causes mutations?

In 1868, Charles Darwin published a new work titled *Variation and Evolution in Plants*.[21] The first volume was replete with examples of the results of artificial selection in producing new animals and plants. In that volume, Darwin was extending and solidifying his theory of evolution beyond all of the examples he had already provided in 1859 in *On the Origin of Species*. But the second volume was a different story. The first three chapters dealt with inheritance, the fourth with the laws of variation, but in the final chapter, Darwin set forth his "Provisional Theory of Pangenesis." Darwin knew that for natural selection to work, existing "paint pot" or blending theories of inheritance would not suffice because the discrete variants upon which natural selection was supposed to act would be swept away in the mixing process. Just think of a few drops of black paint added to a can of white paint. No, Darwin reasoned, the units of heredity must be particulate and favorable units would be selected. Once the notion of particles was accepted, the next question was how they were transmitted and the final question was how variants of these particles arose.

Darwin assumed that his particles, which he called gemmules, were distributed throughout the organism. They were subsequently gathered together from all over the organism to constitute the sexual elements that would lead to the development of the next generation. To create the variation, Darwin needed to have a means of altering gemmules. He proposed two mechanisms. First, he imagined that, if the reproductive organs were injured in some way, gemmules might fail to aggregate properly. There might be too many of some gemmules and too few of other gemmules, resulting in modification and variation. Darwin's second mechanism was Lamarckian. He imagined that gemmules in a particular part of the body could be modified directly in response to changed conditions. But for Darwin, the important thing was that the two mechanisms he envisioned for altering gemmules would create the variation on which natural selection could act.

Herman J. Muller, writing 80 years later, well after the rediscovery of Mendel's principles, knew a lot more about genetics and about the process of mutation. He thought correctly that most mutations were deleterious. "Our Load of Mutations" was the title of the presidential address that he gave at the second annual meeting of the American Society of Human Genetics in 1949.[22] Muller had been a graduate student in Thomas Hunt Morgan's Fly Room at Columbia University, where many of the basic principles of genetics were first worked out using the fruit fly, *Drosophila melanogaster*. In 1926, he proved that X-rays could induce mutations and chromosome rearrangements in fruit flies. For this discovery, he was awarded the Nobel Prize in Physiology or Medicine in 1946.

Muller's assumption was that, with regard to fundamental processes like mutation, the human genome was much like that of a fruit fly. If that were the case, it should constantly be changing due to the process of mutation. A few of these mutations would be beneficial, but most would be deleterious. They might arise "spontaneously," meaning that the mechanism that caused the mutation to occur was unknown, or mutagenic agents such as X-rays might induce them. Over time, these mutations, most of which would be recessive, would accumulate in the population, constituting what Muller called their genetic load. In prehistoric times, a rough equilibrium would have been established in which people made less fit by their genetic load would have been eliminated from the population. But because of advances in such things as sanitation, housing, nutrition, and medicine, Muller speculated that the genetic load in mankind was increasing. Muller then made a distinctly eugenic plea. Those people with high genetic loads, presumably identifiable by genetic disorders, should refrain from reproduction and make way for those "more fortunately endowed to reproduce to more than the average extent."[23]

A number of years after Muller's discovery that X-rays were mutagenic, the first chemical mutagens were found. Charlotte Auerbach was working at the Institute of Animal Genetics at the University of Edinburgh when Muller arrived there in 1938 fresh from the Soviet Union from which he had made a hasty exit.[24] Like many young American intellectuals in the 1920s and '30s, Muller had been enticed by the utopian mirage of a classless society that seemed to shimmer far across the sea following the Bolshevik revolution. He left for the Soviet Union in 1933. In 1935, he published a book titled *Out of the Night*. In his book, Muller advocated the use of artificial insemination as a means for improving the human stock. Muller sent a copy of *Out of the Night* to Stalin. The book seemed to have infuriated Stalin who by then was in the thrall of Trofim Lysenko. Lysenko's essentially neo-Lamarckian ideas, roughly speaking that the proper environment could improve the stock, were much more in keeping with the notion of a classless society than Muller's elitist eugenic notions. Life became unpleasant for Muller in the Soviet Union. So he spent a few weeks in the laboratory of the noted geneticist Boris Ephrussi in Paris before accepting an offer from F. A. E. Crew, Auerbach's boss, to go to Edinburgh.

Muller inspired Auerbach, who now desired to work on mutation. The Edinburgh pharmacologist, A. J. Clark, had noted similarities in the lesions caused by X-rays and mustard gas. Using a clever genetic system devised by Muller for detecting sex-linked lethal mutations in fruit flies, Auerbach found that mustard gas was a mutagen. However, because World War II had started and the British feared the Nazis might use mustard gas against them, Auerbach's paper could not be published until 1946. Thereafter, a host of new chemical mutagens were discovered using various models.

Many different testing systems have been developed for detecting mutagens, but probably the best known of these is the Ames test.[25] This test, developed by Bruce Ames at the University of California, Berkeley, employs mutant strains of the bacterium *Salmonella enterica* sv *typhimurium.* This organism, a close relative of *E. coli,* is best known for causing food poisoning. But *S. enterica* can grow on simple media because it is able to manufacture its own amino acids and vitamins. One of these amino acids is histidine. Ames used several tester strains unable to synthesize histidine that were well characterized as to the mutational defects they carried (e.g., missense mutations, frameshift mutations). Tester strains were then spread over Petri dishes containing media with enough histidine to permit them to grow a little and chemical mutagens were spotted on the dishes in several different places. If the mutagen could revert the original mutant to the wild type or normal condition, a mass of colonies appeared at that point. Some chemicals, called pro-mutagens, are not active in the *Salmonella* tests, but may be active in mammals. Ames was able to test for this possibility by adding a rat liver extract that contained the enzymes required to convert the promutagen to its active form. Of particular significance was the relationship Ames discovered between mutagenic compounds and carcinogens.

In 1775, Sir Percivall Pott, a distinguished surgeon at St. Bartholomew's Hospital in London, made the first connection between chemicals and cancer.[26] Pott observed that chimney sweeps had a high incidence of scrotal cancer, later shown to be squamous cell carcinoma. This resulted in the Chimney Sweeper's Act of 1788. It may be the first piece of environmental legislation ever passed. How effective it was is difficult to assess as its major recommendation was to admonish the boys to go to church on Sunday, but the act also limited a sweeper to six apprentices who must be at least eight years old.[27] This act was not generally enforced, but a much more serious piece of legislation was passed in 1875, which required chimney sweepers to be authorized by the police to carry on their business in a district, thus providing enforcement.

Although Pott's observation made the connection between chemicals and cancer, the link between chemicals, mutation, and cancer was made in 1975 in an important paper from Ames's group titled "Carcinogens are Mutagens: Analysis of 300 chemicals."[28] In addition to specific chemical carcinogens, cigarette smoke condensates and soot were found to be mutagenic. Since Ames reported his results, far more detailed studies have been published in which specific chemicals present in cigarette smoke, soot, diesel exhaust fumes, and so forth have been tested to see whether they were mutagenic in the Ames test. For example, a Swedish study published in 1980 tested 239 of the more abundant chemicals present in tobacco and tobacco smoke for their ability to induce mutations in the Ames test.[29] Only 9 were found to be mutagenic, including a known mutagen benzo(a)pyrene that is found in coal tar, in automobile exhaust fumes (especially from Diesel engines), burnt toast, and cooked meats. In fact, benzo(a)pyrene was very likely one of the culprits, if not the culprit, that caused Pott's chimney sweeps to develop scrotal cancer.

In 1976, Ames and his colleagues published a paper claiming that hair dyes contained a variety of mutagens.[30] There are three kinds of hair-coloring products. Temporary and semipermanent coloring dyes do not penetrate the hair shaft. Permanent dyes do penetrate the hair shaft and last until new growth makes a new application of dye necessary. The dyes studied by Ames were of the latter type. The carcinogenic potential of permanent hair dyes has been the subject of many subsequent studies. It is true that permanent dyes do contain certain compounds that are carcinogenic in high enough concentrations. However, authors of recent reviews have concluded that there is no strong evidence that hair dyes increase the risk of cancer in people who use them, although hairdressers may face an increased risk of developing bladder cancer. It is fortunate that the risk of hair dyes inducing cancer is minimal because the population using these dyes includes about one-third of women in America and Europe and around 10% of men over 40. Furthermore, permanent dyes represent 70% of the market.

In the 1970s, guidelines were being formulated in the United States for testing environmental chemicals.[31] In the 1980s, refinement of the guidelines continued with food additives being included in 1982. It had also become apparent that the Ames test was not all-inclusive. In particular, additional tests were added to check for chromosomal changes. These, of course, could not be carried out in bacteria. The FDA issued further refinements of the testing requirements in 1993. In the late 1990s, the International Conferences on Harmonization of the Toxicological Requirements for Registration of Pharmaceuticals for Human Use achieved consensus. The tests required for pharmaceutical registration would include the Ames test and tests for chromosome damage.

We are, of course, exposed to a potent mutagen every day that is not a chemical: ultraviolet light. Most weather reports today, especially in the summer, come complete with a UV Index, developed by the National Weather Service and the Environmental Protection Agency.[32] The strength of UV radiation is given on a scale of 1 (low) to 11+ (extremely high). The National Weather Service calculates the predicted UV Index for the next day in each area of the United States. UV light is part of the electromagnetic spectrum. The visible spectrum extends from 700 to 400 nm (nanometers or billionths of a meter) followed in decreasing wavelength by UV. UV light is arbitrarily divided into three sets of wavelengths. UVA (400–320 nm), UVB (320–290 nm), and UVC (280–100 nm).[33] UVC is absorbed by the ozone layer, but UVA and UVB penetrate the atmosphere and your skin.

UV radiation is considered to be the main cause of nonmelanoma skin cancers, including basal cell carcinoma and squamous cell carcinoma.[34] Basal cell carcinomas are estimated to strike over a million people in the United States every year, whereas squamous cell carcinomas affect around 250,000. UV also often plays a key role in melanoma formation, the most lethal form of skin cancer. UVA penetrates deeper into the skin than UVB and is largely responsible for skin aging and wrinkling. UVA also plays a major role in tanning and so the light in tanning booths is principally UVA. The high-pressure sunlamps used in tanning salons emit 12 times the UVA of sunlight so, unsurprisingly, people who make use of tanning salons are 2.5 times more likely to develop squamous cell carcinoma and 1.5 times more likely to elaborate basal cell carcinomas. UVB is the chief culprit in producing the reddening of the skin in sunburn.

The FDA currently approves 17 different compounds for filtering UVA and UVB light. With a few exceptions, they either filter one set of frequencies or the other. Because UVB produces sunburn, sunscreens are currently classified according to their ability to protect against UVB. Thus, a sun protection factor (SPF) of 15 means that it will take 15 times as long for the skin to achieve the level of redness it would without the sunscreen. An SPF15 sunscreen blocks out 93% of the UVB rays, SPF 30 protects against 97%, and SPF 50 does the job for 98%.

The trouble with the SPF system is that it does not take UVA into account. In addition, although UVA tans, it also causes skin cancers. Within the next few years, the FDA will probably come up with guidelines for manufacturers to follow that indicate the degree of protection of a given sunscreen against UVA as well as UVB. Obviously, if you are interested in getting a nice tan, you will probably choose a sunscreen that protects well against UVB, but not against UVA. It's a Catch-22.

Because mutations are usually detrimental, organisms from single-celled bacteria to humans have developed repair systems for DNA that recognize and correct mutational lesions. DNA damage by ultraviolet light is repaired in part by an enzyme called a photolyase that requires visible light to work. But photoreactivation is just one of a whole collection of processes by which cellular DNA damage is identified and corrected (see Repair in Glossary for more details).

Some mutations are worse than others: variation in the cystic fibrosis gene

It has been known for centuries that the infant who tastes salty when kissed is marked for death. Such infants were thought to be bewitched.[35] A salty taste is the hallmark of cystic fibrosis. Cystic fibrosis is a relatively common ailment. About 8 million Americans possess a copy of the defective gene, which is recessive, whereas about 30,000 children and young adults have the disease.[36] Around 1,000 babies are born each year in the United States with cystic fibrosis. Cystic fibrosis is not randomly distributed among different racial or ethnic groups, but is instead found mostly within the Caucasian population (see Chapter 3, "Ethnicity and genetic disease").

A single gene defect characteristically signals its presence by causing a variety of symptoms.[37] Depending on the nature of the mutation, people with cystic fibrosis either have a defective or missing protein called CFTR. This protein channels chloride ions out of mucus-producing epithelial cells that line the surfaces of parts of the body like the lungs and the intestines as well as tissues that produce sweat, saliva, and digestive enzymes. Water normally follows the chloride ions out of these cells by the process of osmosis. In cystic fibrosis, chloride transport is blocked so water does not pass out of the cells. The mucus becomes much more viscous than normal because there is insufficient water available to dilute it.

The external surfaces of many of the cells, such as those that line the lungs, are covered with beating cilia that propel the mucus, and any bacteria that may be contained in it, into the throat and out of the body. In cystic fibrosis, the viscous mucus that accumulates prevents the cilia from beating. Clearing the lungs daily of mucus is essential and requires another person to clap the sufferer's back vigorously. Because the accumulating mucus prevents the cilia from clearing out bacteria from the lungs, cystic fibrosis victims are also susceptible to bacterial infections by organisms like *Pseudomonas aeroginosa* and *Burkholderia cepacia.* These are known as opportunistic pathogens because they do not normally cause disease in healthy people or tissues. Despite these manifest complications, survival rates for cystic fibrosis sufferers have improved dramatically. As a careful study carried out in northern Italy reports, death from cystic fibrosis was expected to occur in the first decade of life 40 years ago, but today the median survival rate is 37.7 years among the patients studied.[38] This survival rate is somewhat higher than in the United States.

Since 1989, when Francis Collins and his colleagues isolated and sequenced the cystic fibrosis (CF) gene, more than 1,500 mutations have been identified in the gene that can cause cystic fibrosis. A mutation called phe508del or δ508 causes a severe form of the disease, and accounts for 70% to 90% of the total cases in most Caucasian populations.[39] This mutation results in omission of phenylalanine at position 508 in the CFTR protein chain because of the deletion of the three base pairs encoding this amino acid at the corresponding position in the CF gene. The mutant protein does not fold properly and is destroyed. This mutation has been around for at least two millennia having been identified in the molars of 3 of 32 Iron Age Austrians who were found buried in cemeteries near Vienna.[40]

Individuals with mutations like phe508del that result in complete loss of CFTR function have the most severe symptoms of cystic fibrosis. Others have mutations that allow partial functioning of the protein and their symptoms are less severe. In fact, the following functional classification has been developed for cystic fibrosis mutations: Class I, defective production of the CFTR protein; Class II, defective protein processing; Class III, defective protein regulation; Class IV, defective conductance of chloride ions by the CFTR protein; and Class V, reduced amounts of functional CFTR protein. Classes I–III are associated with the severe forms of the disease, whereas individuals in Classes IV and V have less-severe cases of cystic fibrosis.[41] So, as the case of cystic fibrosis shows, the type of mutation determines the nature of the defect and the severity of the disease. The same is true of other disease genes as well (e.g., those encoding hemoglobin) where more than one mutation is known.

When chromosomes foul up

John Langdon Down, a London doctor, was appointed resident physician at the Earlswood Asylum for Idiots in Surrey in 1858.[42] He was subsequently made medical superintendent of the institution. During his years at the asylum, he set up a system that became a model for care of the mentally ill in the United Kingdom. In 1866, Down published a paper titled "Observations on an Ethnic Classification of Idiots."[43] His purpose was to devise a system for categorizing the "feeble-minded, by arranging them around various ethnic standards." Down's intent was simply to make it easier to classify and treat different kinds of mental disease based on the physical appearance of the individual. He based his system on observations made both at Earlswood and at the London Hospital. He noted that there were many examples of "the great Caucasian family," but also some of the "Ethiopian variety." These people did not have dark skin pigmentation, but had woolly hair and puffy lips. Others "arrange themselves around the Malay variety."

But the main focus of Down's paper was on what he called "the great Mongolian family." He reported that the "Mongolian type of idiocy occurs in more than ten percent of the cases which are presented to me." Down then went on to describe the characteristics of people with the syndrome that today bears his name. And at the very end of his paper, Down used the examples he had presented to argue against the notion that the "great racial divisions are fixed and definite." If they were, "how comes it that disease is able to break down the barrier, and to simulate so closely the features of the members of another division. I cannot but think that the observations which I have recorded, are indications that the differences in the races are not specific but variable."

Terms like "Mongolian Idiocy" and "Mongolism" continued to be used well into the twentieth century. However, Dr. P.W. Hunter at the Royal Albert Asylum in Lancaster took Down's hypothesis a step further. He suggested that Down syndrome did not result from reversion to a more primitive race, but approached more closely primate relatives of humans such as the gorilla or chimpanzee.[44] Francis Crookshank, a London physician, embraced and modified this theme. He hypothesized that children whose features resembled those of more "primitive" races had never undergone complete development in the womb. These children were more like our primate ancestors than even the "inferior" Mongol race. In short, the Mongolian idiot was a sort of missing link. Crookshank elaborated these ideas in a popular book titled *The Mongol in Our Midst* that was first published in 1924 and came out in two subsequent editions in the early 1930s. This, of course, was a time when the eugenics movement was reaching its zenith. The idea of linking "feeblemindedness" with the idea of an "inferior race" and even with our primate ancestors probably had considerable appeal.

In the first part of the twentieth century, "feeblemindedness" was the catchall label for mental disease. In the early 1930s, Lionel Penrose, an English physician, began to make sense out of the many forms of mental disease.[45] As the result of a survey conducted by Edmund O. Lewis, a British physician and expert on mental health, for the government's Joint Committee on Mental Deficiency, it had become clear that the various grades of mental defectives needed to be sorted out and properly categorized. The Royal Eastern Institution at Colchester seemed the ideal place for this work. There were more than a thousand patients there and it was in a part of England where Lewis's survey had detected the greatest number of mentally deficient children.

Funds were found to support the project and Penrose was appointed as the investigator in October 1930. He soon realized that there were many different types of mental retardation and he began to sort them out. By the time the third edition of Crookshank's book came out in 1931, Penrose had a group of 42 "Mongolian imbeciles" at Colchester and was searching for others in the nearby London hospitals. He quickly concluded that Mongolian "imbeciles are no more racially Mongolian than any other imbeciles." Penrose appreciated them for their gentle childlike quality and began to refer to them as Down syndrome patients. Although it had been noted early in the twentieth century that the frequency of Down syndrome births increased with maternal age, it was Penrose who established this relationship convincingly. Older mothers were much more likely to have Down children than younger mothers.

But what was the basis of this strange disease whose incidence increased with maternal age? We know today that the great majority of Down syndrome cases arise in people who have a third copy of chromosome 21 (trisomy 21), but at the time Penrose was conducting the Colchester survey, knowledge of human chromosome number was primitive. The problem was that there were a lot of chromosomes and they were hard to see.[46] Early attempts to count human chromosomes gave an estimate of around 24. But these estimates made use of material from corpses, often of executed criminals, and the chromosomes tended to clump together soon after death. The Belgian cytologist Hans von Winiwarter was familiar with the clumping problem and in 1912 came up with an estimate of 48 using fresh tissue from female oogonia and 47 chromosomes from male spermatogonia. He missed the Y chromosome in the male tissue, concluding that males were of the XO chromosome type as were some species of insects.

Despite von Winiwarter's work, most investigators continued to assume that the correct human chromosome number was 24. That all changed in the early 1920s—thanks to the work of Theophilus Painter, a gifted cytogeneticist at the University of Texas. One of Painter's former students, practicing medicine at a state mental institution in Austin, provided Painter with the testes of one white and two black men. Inmates of the institution were occasionally castrated for "therapeutic reasons" because "excessive abuse coupled with certain phases of insanity made the removal of the sex glands desirable." Whether or not this rational made sense, Painter had fresh material and immediately went to work. He confirmed von Winiwarter's number and added the Y chromosome, the smallest human chromosome, which von Winiwarter had missed.

Painter's count of 48 chromosomes stood until 1955 when Joe Hin Tjio, working in the laboratory of the Swedish geneticist Albert Levan at the University of Lund, counted only 46 chromosomes while examining human lung cells in tissue culture. Tjio and Levan published this number, which is correct, in 1956. They were able to visualize the chromosomes much better than their predecessors due to several important technical advances in human chromosome cytology. Over the years, methods for banding chromosomes into light and dark regions have also been developed.[47] Quinacrine mustard was the first dye used for this purpose (Q-banding). This is a fluorescent dye that causes a chromosome to appear as a mosaic of bright and dull bands. The bright bands are rich in adenine and thymine while in the dull bands guanine and cytosine predominate. Giemsa stain is the most commonly used banding dye (G-banding). Some pretreatment, for example with a protein-digesting enzyme is required so that the chromosomal DNA becomes accessible. Like quinacrine, this dye also stains adenine-thymine rich regions dark and regions where guanine and cytosine predominate lightly.

The man who discovered the chromosomal basis of Down syndrome was a French pediatrician and geneticist named Jérôme Lejeune.[48] In 1959, Lejeune, who was working in Raymond Turpin's laboratory at the Hospital St. Louis in Paris, and his colleague Marthe Gautier reported that Down syndrome patients had an extra copy of chromosome 21. Lejeune's discovery of the extra chromosome 21 in Down syndrome was ironic for him. Lejeune was a pro-life Catholic who specialized in the treatment of children with Down syndrome. Most women who find that their fetus has an extra chromosome 21 choose to abort it.

How does the fetus acquire an extra chromosome? To answer this question, a short detour into the physiology of human egg formation is in order. By the time the fetus is 20 weeks old, diploid oogonia have given rise to all of the primary oocytes a female fetus will ever possess. These oocytes are arrested in the first meiotic division. No further change occurs until many years later when a girl's menstrual cycle begins. Usually, once a month some of the primary oocytes enlarge and complete the first meiotic division. Each primary oocyte produces a secondary oocyte and a polar body that is discarded.

Following fertilization, the secondary oocyte undergoes the second meiotic division to yield the fertilized egg and a second polar body that is discarded.

If an error occurs at the first meiotic division so that both copies of chromosome 21 go into the secondary oocyte instead of one going into the first polar body and the second meiotic division occurs normally, the secondary oocyte will end up with two copies of chromosome 21 (primary nondisjunction). An error can also occur at the second meiotic division so that the egg ends up with two copies of chromosome 21 (secondary nondisjunction). In either case, the fertilized egg will contain three copies of chromosome 21. If the resulting embryo is not aborted, a Down baby will result. The probability of a nondisjunctional event resulting in a Down infant increases dramatically with maternal age (see Table 2–3). The presence of an extra copy of chromosome 21 accounts for about 92% of all Down syndrome births, but in 3% to 5%, the extra copy is missing. In these cases, a piece of chromosome 21 becomes attached to chromosome 14. Unlike the majority of Down syndrome cases, this type of translocational Down syndrome can be inherited, but hereditary transmission occurs in only one out of three cases.[49]

Techniques like chorionic villus sampling, amniocentesis, and sampling of blood from the umbilical cord of the fetus (see Glossary) can be used to determine whether an expectant mother will give birth to a Down syndrome baby. A study published in 1999 that summarized studies of termination rates for several diseases including Down syndrome reported that in the United States anywhere from 87% to 98% of Down syndrome pregnancies are terminated.[50] Similar rates of termination were observed in other countries. Termination rates so high understandably shocked *Washington Post* columnist George Will whose son Jon has Down syndrome. In a 2005 opinion piece titled, "Eugenics By Abortion,"[51] Will pointed out that his son readily navigated the Washington subway system to attend Wizards basketball games and soon would be attending Washington Nationals baseball games.

Table 2–3 Relationship of Down syndrome to maternal age

Maternal age	Incidence of Down syndrome/1,000 births
20	0.579
25	0.799
30	0.795
32	1.044
34	2.098
36	2.582
38	5.113
40	12.188
42	17.751

Modified from Hook, Ernest B. and Agneta Linsjö. Down Syndrome in Live Births by Single Year Maternal Age Interval in a Swedish Study: Comparison with Results from a New York State Study. *American Journal of Human Genetics* 30(1978): 19–27.

Will's son Jon evidently copes pretty well. This illustrates the fact that mental impairment can be quite mild (IQ: 50–70), but it can also be moderate (IQ: 35–50) and is occasionally severe (IQ: 20–35).[52] Unfortunately, there is no way to predict the degree to which a child with Down syndrome will be mentally impaired before birth. There can be other complications as well. For instance, about 75% of Down children show some degree of hearing loss, and obstructive sleep apnea affects 50% to 75% of individuals with this syndrome. These considerations, especially the unpredictability of the degree of mental impairment, probably explain why such a high proportion of prospective parents opt for termination of pregnancy when told they are going to have a baby with Down syndrome. The Committee on Genetics of the American Academy of Pediatrics has issued specific guidelines for the care of children with Down syndrome.[53]

Remarkably few children with chromosome anomalies are born because a highly effective natural system exists for their detection.[54] The vast majority (90%) of chromosomally abnormal pregnancies miscarry in the first trimester of pregnancy, whereas 93% of those with normal chromosome complements continue. Recurrent miscarriage, defined as three consecutive pregnancy losses before week 24 of pregnancy, occurs in 0.5% to 1.0% in all fertile couples of reproductive age. According to one hypothesis, recurrent miscarriage results from the failure of nature's control system, allowing "poor quality" embryos to implant and then miscarry.

Chromosome 21 is the smallest chromosome other than the Y chromosome. About 1 in 29,000 babies are born with trisomy 13 (Patau syndrome) and 1 in 7,000 with trisomy 18 (Edwards syndrome).[55] Most affected babies die before there first birthday. Around 50% of spontaneous abortions before 15 weeks of gestation result from chromosomal imbalances with trisomies accounting for 50% of these events. The incidence of trisomy has been studied in some detail for several other chromosomes. As is true for chromosome 21, the vast majority of nondisjunction events leading to trisomy are maternal in origin for chromosomes 13, 15, 16, and 18. Trisomy 16 is the most common human trisomy and its frequency clearly increases with maternal age. However, nondisjunction events appear to be independent of maternal age in the case of several other chromosomes.[56]

Tortoiseshell cats and sex chromosomes

Tortoiseshell cats are rather odd. The males are either black or orange, but the females have patches of both black and orange fur. Why is this? The sex chromosomes of cats are like ours. Females have two X chromosomes and males have an X and a Y. The primary coat color gene for cats is located on the X chromosome. It comes in two forms. One codes for black color and the other codes for orange color. Ergo, males are either black or orange. But why do the females have patches of orange and black? The explanation is a phenomenon called X-chromosome inactivation, first postulated by Mary Lyon in 1961.[57] Early in the development of the female cat, one X chromosome or the other is turned off. In a cell line where the X chromosome having the orange form of the gene is turned off, a patch of black fur will appear. Conversely, in a cell line where the X chromosome having the black form of the gene is turned off, a patch of orange fur will result.

The phenomenon of X-chromosome inactivation also occurs in women. In 1949, Murray Barr, a Canadian physician and cytologist, discovered a small structure in the nuclei of female cells that stained deeply.[58] Male cells lacked this structure that was subsequently named the "Barr body." Meanwhile, Paul Polani, a doctor at Guy's Hospital in London, who was investigating congenital heart disease, came across three women with a peculiar condition found exclusively in women called Turner's syndrome. These women had an aortal defect, usually found in men, and a webbed neck. They did have female genitalia, but very poorly developed ovaries. Polani wondered whether the Turner's women were really women after all. Examination of their cells for the presence of the Barr body revealed that it was missing, so genetically, these women were behaving like males, but they lacked a Y chromosome. That is, they were XO.

The most characteristic feature of men with Klinefelter's syndrome is that they have very small testes and are infertile.[59] Klinefelter's men have two X chromosomes plus a Y chromosome (XXY) 80% of the time and exhibit a Barr body. The other 20% of Klinefelter's individuals have additional X chromosomes (e.g., XXXY, XXXXY) and sometimes an extra Y chromosome (e.g., XXYY) as well. The number of Barr bodies is always one less than the number of X chromosomes. That is, all but one of the X chromosomes is inactive.

In the summer of 2009, the rare birth of a male tortoiseshell cat in England made it into the national press.[60] The eight-week-old cat was brought to Mrs. Karen Horne, a veterinarian, in Harpenden, Hertfordshire, who adopted it. She had never seen such an animal nor had her colleagues at the veterinary clinic with a collective 30 years of experience. Horne named the cat Eddie after Eddie Izzard, the stand-up British comedian who frequently cross-dresses because Eddie the cat "is essentially a boy dressed in girls' clothing." Eddie is a Klinefelter's cat. The fact that he is a tortoiseshell shows that he is a mosaic like a female tortoiseshell cat with one X chromosome expressed in some tissues and the second X chromosome in other tissues.

X chromosome inactivation functions as a dosage-compensation system. It ensures that the dosage of genes on the X chromosome is the same in both sexes. But because both X chromosomes are expressed separately in different tissues of the female, disease-causing mutations on the X chromosome (e.g., those causing hemophilia A and B) are masked in women. Dosage compensation is achieved by the X inactivation center (*Xic*) located on the X chromosome.[61] *Xic* is somehow able to count the number of X chromosomes. *Xic* contains a gene called *Xist* whose product is an RNA molecule whose sole role is to silence the X chromosome to be inactivated (*Xi*), coating it with this RNA. On the active X chromosome (*Xa*), this process does not occur. Exactly how chromosome counting is done, how one X chromosome or the other is chosen for inactivation, and how the inactive state is maintained are questions for the future.

X inactivation is a form of what is called epigenetic gene silencing, but this phenomenon is not restricted to the X chromosome. Consider Angelman syndrome, for example. In 1965, Harry Angelman, an English pediatrician, published a paper titled "'Puppet' Children. A report of three cases."[62] All three children walked stiffly with a jerky gait, did not speak, laughed frequently, and were subject to seizures. At first, Angelman did not know how to describe the children, but, while vacationing in Italy, he visited the Castelvecchio Museum in Verona. There he saw a painting by Giovanni Francesco Caroto that was titled at the time "Boy with a Puppet," but is now called "Boy with a Drawing."

"The boy's laughing face and the fact that my patients exhibited jerky movements gave me the idea of writing an article about the three children with a title of Puppet Children. It was not a name that pleased all parents but it served as a means of combining the three little patients into a single group. Later the name was changed to Angelman syndrome."[63]

Angelman syndrome most commonly arises as the result of the deletion of part of the maternal copy of chromosome 15, containing a gene called *UBE3A*. This gene is normally present in both the paternal and maternal copies of chromosome 15. Both copies are active in most tissues.[64] However, in certain areas of the brain, notably the hippocampus and the cerebellum, only the maternal copy is functional because the paternal copy has been imprinted (silenced) through the addition of specific chemical (in this case, methyl) groups. The absence of expression of the *UBE3A* gene in these regions of the brain appears to be responsible for causing the various symptoms of the disease.

What happens if there is a similar deletion in the paternal copy of chromosome 15? A different genetic disease results called Prader-Willi syndrome. People with this disease are short, have small hands and feet, and have mild mental retardation.[65] They may also be subject to temper tantrums, stubbornness, and compulsive behavior. A major problem that accompanies Prader-Willi syndrome is an insatiable appetite so someone with this disease will rapidly become obese unless his or her diet is strictly controlled. The paternal region of chromosome 15 contains several genes that are expressed exclusively from this chromosome with the same genes in the maternal region being inactivated.[66] However, as is true in a normal individual, the maternal *UBE3A* gene is unaffected and functional in a person with Prader-Willi syndrome.

People with either of these syndromes often have unusually pale skins and fair hair. The reason is that the deletions in both cases often include the *OCA2* gene. *OCA2* encodes the P protein.[67] The P protein is found in the cells (melanocytes) that specialize in producing the dark pigment melanin. When only one copy of *OCA2* is present, less P protein is made. Hence, the amount of melanin is reduced causing the skin and hair to lighten.

The kind of epigenetic silencing seen for the X chromosome and in the case of Prader-Willi syndrome is part of the permanent human epigenetic regulatory system. Epigenomics, a new and important area of research, focuses on epigenetic alterations across the entire genome. Some of these epigenetic modifications are part of the normal regulatory apparatus, but others are not, although they may persist for many generations. These latter sorts of alterations are often triggered by some environmental cue and may involve addition of methyl groups to specific genes that usually silences them.

On January 6, 2010, *Time* magazine ran a lead article by John Cloud titled "Why Your DNA Isn't Your Destiny."[68] Its subject was the emerging science of epigenomics and it reviewed in some detail the work of Dr. Lars Olov Bygren, a preventive-health specialist at the Karolinska Institute in Stockholm. Dr. Bygren's research focused on the population of Sweden's most northern county, Norrbotten. During the nineteenth century, the harvests there were very unpredictable. Some years, 1800 or 1836 for instance, saw total crop failure while in other years, 1801 and 1844 for example, crops were abundant so people who had gone hungry the year before were able to eat as much as they wanted.

Bygren zeroed in on a random sample of 99 individuals born in the Overkalix parish of Norbotten in 1905. He then used historical records to follow their parents and grandparents back to their births. By analyzing the detailed agricultural records that were available, Bygren and two colleagues could calculate how much food had been available to the parents and grandparents when they were young. Their astonishing finding was that girls and boys who were fortunate enough to experience winters when food was abundant produced children and grandchildren with strikingly shorter life spans than the sons and grandsons of those who had endured winters marked by poor harvests. As Bygren recognized, the shortened life span of the children and grandchildren of those who had eaten well must have been the consequence of some sort of epigenetic modification whose basis has yet to be determined. How long these variations in life span will persist only time can tell.

3

Ethnicity and genetic disease

Some ethnic groups have higher frequencies of certain genetic diseases than others. There are two main reasons. The first involves selection. Sometimes a disease gene can actually confer a selective advantage on its carrier. The classic case involves sickle cell anemia. Carriers of the mutant gene are more resistant to the ravages of malaria than people who lack the gene. The fact that African Americans are far more likely to have sickle cell anemia than other ethnic groups reflects the fact that their ancestors came from parts of Africa where malaria is rampant.

The second reason why disease genes are at a higher frequency in some populations than others has to do with chance alone (see Chapter 1, "Hunting for disease genes"). The so-called "Jewish genetic diseases," including Tay-Sachs disease, may well fall into this category, but a stronger case can be made for the two founding populations of French Canada, one in Quebec and the other in Acadia.

The potential protective role that disease genes might have against threats like malaria was first suggested by John Burdon Sanderson Haldane. "J. B. S" Haldane, as he preferred to be called, was a big man with a walrus mustache.[1] He was the son of an Oxford professor of physiology and descended from an ancient line of Scottish aristocrats. Following Eton, which he hated, and New College, Oxford, Haldane volunteered for the Scottish Black Watch in World War I. He was immediately sent to the front, serving initially as the bombing officer for the third battalion before becoming trench mortar officer for the first. Much to his dismay, he found that he actually enjoyed killing Germans. He was wounded twice and personally delivered bombs and conducted sabotage behind enemy lines. His commander called him "the bravest and dirtiest soldier in my army." Fortunately, Haldane survived the war to become one of the greatest population geneticists of all time, spending most of his career at University College in London. In the 1920s, he became interested in studying the mechanisms involved in the process of natural selection. In his 1932 book *The Causes of Evolution*, he pointed out that one "of the principal characters possessing survival value is immunity to disease." This idea was initially ignored until mounting evidence began to indicate that genes conferring disease resistance did in fact exist.

In 1949, Haldane participated in a symposium on ecological and genetic factors in animal speciation held at the Istituto Sieroterapico Milanese in Milan.[2] In his paper, he emphasized once again the decisive role that disease and genetic resistance to disease could play in evolution. After Haldane's presentation, the Italian geneticist Giuseppe Montalenti mentioned a conversation he had previously had with Haldane in which Haldane suggested that carriers of the thalassemia gene might be more resistant to malaria than normal individuals. This meant that carriers of a serious genetic disease would be at a selective advantage to normal individuals in parts of the world where malaria is endemic even though one-quarter of the children produced by two carriers would become ill with thalassemia.

Malaria and sickle cell disease

The Ga, Fante, and Ewe people live mostly in the Republic of Ghana.[3] There are many different languages in Ghana and each of these tribal groupings speaks a different tongue. In fact, the Ga form a tribal group with the Adangbe. Today, the Adangbe are found to the east and the Ga groups to the west. Their languages are now mutually unintelligible despite deriving from a common proto-Ga-Adangbe ancestral language. These and other West African tribal groups have known about sickle cell disease for hundreds of years and each language has its own word for the syndrome. It is, for example, *chwechweechwe* in Ga, *ahotutuo* in Twi, a language belonging to the larger Akan language group and widely spoken in Ghana, *nuidudui* in Ewe, and *nwiiwii* in Fante. All these names contain repeating syllables possibly symbolizing the repeating painful episodes caused by the disease or perhaps in imitation of the cries and moans of those suffering from the disease.

Most of these tribes were probably in place when the Portuguese arrived in 1471.[4] Because Ghana was known to be the source of gold that reached Muslim North Africa via Saharan trade routes, the area became known as the Gold Coast. As European plantations began to sprout up in the Americas in the 1500s, slaves soon dwarfed gold as the principal export of the region. The slave trade reached its peak in the eighteenth century. It has been estimated that 6.3 million slaves were shipped from West Africa to North and South America and the West Indies between 1500 and the end of the slave trade. About 4.5 million individuals were sent abroad between 1701 and 1810 alone. Not only did African slaves travel west to the New World, but sickle cell anemia accompanied them. Today, about 1 in 500 African Americans develop sickle cell anemia as compared with 1 in 100 in Africa. The difference in frequency is most easily explained by the fact that malaria is no longer acting as a selective agent in the United States.

The reason why sickle cell anemia is so prevalent in African populations started to emerge in 1946 when E. A. Beet, a doctor in northern Rhodesia (now Zambia), observed that among a population of patients in his hospital, 15.3% whose blood was normal had malaria, but only 9.8% of those who were carriers of the genetic mutation that causes sickle cell anemia (sickle cell trait) were suffering from the disease.[5] Anthony Allison, who had taken Oxford degrees in biochemistry and genetics just after World War II, began a careful examination of the situation in Africa in the early 1950s. In 1954, he published a hypothesis as to why sickle cell anemia is so common in much of Africa.[6] He had found that in some tribes, up to 40% of the people had sickle cell trait. This is a much higher frequency than one would expect for such a deleterious mutation. Allison suggested that individuals with sickle cell trait must be more resistant to malaria than normal individuals. This would explain why the mutant gene was in an abnormally high frequency in the population even though the consequences of sickle cell anemia itself were so severe.

Allison also noted that the African regions in which the frequency of sickle cell trait and sickle cell anemia were the highest corresponded in geography to the distribution of the most deadly form of the malaria parasite (*Plasmodium falciparum*). Allison recognized that the sickle cell gene was being maintained as a "balanced polymorphism." That is, carriers with one copy of the normal gene and one copy of the mutant gene (heterozygotes) had a selective advantage over individuals with two copies of the normal gene or two copies of the mutant gene (homozygotes).

Following publication of his paper, Allison was invited to give a presentation of his work at University College London. Afterward, there was a friendly discussion during which Haldane said that he had earlier realized that carriers of the thalassemia gene might be at a selective advantage with regard to malaria, a point that had come out in his 1949 symposium paper. However, Haldane graciously added that speculation was one thing, but that Allison had produced the first real evidence that natural selection operates in humans.[7]

The reason why individuals with sickle cell trait are more resistant to infection by the malaria parasite became apparent with the characterization of the disease. In 1910, Dr. James Herrick, a Chicago physician, described a young black man from Grenada who had a case of severe anemia as having sickle-shaped red blood corpuscles.[8] Once sickle cell trait had been clearly distinguished from sickle cell anemia, it was found that the blood corpuscles of individuals with sickle cell trait did not sickle to any great degree except at very low oxygen tensions. Exactly how sickle cell trait confers malaria resistance is still not fully understood, but it undoubtedly involves an enhanced reaction of the immune system to the parasites as well as destruction of the infected cells by phagocytes.[9]

Malaria is estimated to cause around one million deaths per year with most of these deaths occurring in African children.[10] The immune system helps a person to resist attacks by *P. falciparum* and maternal antibodies transmitted to newborns prior to birth provide some immunity to malaria for the first few months. Thereafter, the child's immune system starts to function. The protective effect of possessing sickle cell trait appears to be the bridge between these two processes, a time at which the child is vulnerable to the most serious consequences of malaria.

Sickle cell anemia was the first genetic disease whose molecular basis was fully understood.[11] Because the disease alters the properties of adult hemoglobin (there are embryonic and fetal hemoglobins too), it is important to understand a bit about the structure of the molecule. An adult hemoglobin molecule is composed of four polypeptide chains called globins. There are two identical alpha (α) globin chains and two identical beta (β) globin chains. The structure of adult hemoglobin is, therefore, ($\alpha_2\beta_2$). Each of the four globin molecules has an iron-containing heme molecule tightly associated with it.

It is the iron atom that carries oxygen to the tissues. The α globin chains are encoded by the *HBA* gene of which there are two copies located near the tip of the short arm of chromosome 16 while the β globin chains are encoded by the *HBB* gene of which there is a single copy that is found on chromosome 11.

Using a technique called electrophoresis, Linus Pauling reported in 1949 that normal hemoglobin (HbA) moved at a different position in an electrical field than sickle cell hemoglobin (HbS). In the mid 1950s, Vernon Ingram was able to show that the reason for this was that the normal beta chain of hemoglobin contained a water-soluble, acidic amino acid (glutamic acid) at position six near the beginning of the protein chain (amino terminus), whereas HbS possessed a water-insoluble neutral amino acid at this position (valine).

This seemingly minor change has profound effects in individuals with sickle cell disease.[12] Not only do their red blood cells sickle severely, but numerous secondary complications result. Among these, two dominate. One is the anemia itself. This results because the cells achieve a normal configuration upon oxygenation following which they sickle as they become deoxygenated. Repeated cycles of reoxygenation and deoxygenation result in permanent damage to the cells and their subsequent destruction.

The other complication arises because the sickle-shaped erythrocytes tend to accumulate within blood vessels, resulting in oxygen deprivation and damage to surrounding tissues. This results in the periodic onset of "painful episodes." Most of these are of mild to moderate severity and can be treated at home. The anemia itself, of course, contributes to overall tissue deprivation for oxygen. Mortality statistics indicate that the first five years of life are the most critical.

Today, over 85% of individuals with sickle cell anemia survive beyond the age of 20, but most will die in their mid-40s.[13] Penicillin treatment of children prevents 80% of life-threatening pneumonia episodes.[14]

The only approved medication for sickle cell anemia is hydroxyurea. This compound works by increasing the proportion of fetal hemoglobin in adults. In fetal hemoglobin, the two β polypeptides are replaced by two gamma (γ) polypeptides so the structure of this hemoglobin is $\alpha_2\gamma_2$. The two gene copies encoding the γ polypeptides are found on chromosome 11 and are distinct from the *HBB* gene. Hence, fetal hemoglobin is normal in a fetus with sickle cell anemia. However, after birth, adult hemoglobin gradually replaces fetal hemoglobin. Hydroxyurea boosts the proportion of fetal hemoglobin in the blood in children from 5%–10% to 15%–20%.[15] Clinical evidence shows that hydroxyurea effects a marked reduction in the number of painful episodes caused by sickle cell anemia and, in one study, reduced mortality by 40%.[16] Exactly how hydroxyurea causes an increase in the proportion of fetal hemoglobin in the blood is not entirely understood, but the drug appears to reduce the production of red blood cells containing sickle cell hemoglobin while favoring the production of cells having a high proportion of fetal hemoglobin.[17]

A 1970 paper in the *New England Journal of Medicine* reported the sudden deaths of four African American army recruits undergoing basic combat training at an army post whose altitude was given as 4,060 feet.[18] It turned out that each of the recruits had sickle cell trait and that the post was Fort Bliss in El Paso, Texas. The resulting fear that individuals with sickle cell trait might be unusually susceptible to the effects of high altitudes led the Air Force Academy to prohibit their entry.[19] In addition, several major airlines mandated that employees with sickle cell trait would be restricted to ground jobs. Furthermore, insurance companies often began to charge higher premiums to people with sickle cell trait. Because African Americans were the predominant group affected by sickle cell disease, these various decisions seemed to smack of discrimination. Consequently, in 1981, the Air Force Academy ended its prohibition against admitting students with sickle cell trait.

In 1972, the preamble to the National Sickle Cell Anemia Control Act erroneously stated that two million Americans suffered from sickle cell "disease." Instead, the preamble should have said that this number of Americans had sickle cell trait because fewer than one hundred thousand had the disease.[20] Despite this error in its preamble, the National Sickle Cell Anemia Control Act has had salutary consequences. The law provided for the establishment of voluntary sickle cell anemia screening and counseling programs; information and education programs for health professionals and the public; and research and training in the diagnosis, treatment, and control of sickle cell anemia. However, early screening programs were fraught with problems, including making the distinction between sickle cell anemia and sickle cell trait. Marked improvement has occurred since then. Most states require newborn screening for sickle cell disease and sickle cell trait as well as a group of other inherited diseases (for a list by state, see reference 21). By its 30th anniversary in 2002, the National Sickle Cell Anemia Control Act had dispensed over $923 million for the various activities mandated by the act via the National Heart, Lung, and Blood Institute of the National Institutes of Health.[22]

African American players dominate college and professional sports like basketball, football, and track. In 1973, a star quarterback for the Coral Gables High School named Polie Poitier was recruited to play football at the University of Colorado at Boulder by its great coach Eddie Crowder.[23] That was Crowder's last year as coach so the new coach of the football team, Bill Mallory, inherited Poitier whom he felt would fit right into the team's lineup. In August 1974, just before the football season began, Poitier went out on a training run. During the run, he collapsed and died. As it turned out, Poitier had also collapsed on the first day of practice in 1973. Poitier's death was the first recorded by the NCAA caused by complications related to sickle cell trait. It also fit a pattern that has become familiar. Poitier's death occurred during conditioning rather than practicing for or playing a game.

The problem of accidental death of individuals with sickle cell trait during training became serious enough for the National Athletic Trainers' Association to issue a consensus statement on the subject in 2007.[24] They note that rapid breakdown of muscular tissue starved of blood (acute rhabdomyolysis) tied to sickle cell trait is the third most important killer of high school and college athletes after cardiovascular conditions and heatstroke. A 2006 survey of NCAA Division I-A schools found that, among schools that responded, 64% screened for sickle cell trait. The consensus statement recommends that no athlete with sickle cell trait should be disqualified from playing his or her sport, but lists a series of precautions that should be observed. The Trainers' Association also encourages all athletes potentially at risk to have a simple and inexpensive blood test to see whether they are carriers of the sickle cell gene.

Sickle cell trait has posed a problem in professional sports too. Just before the Pittsburgh Steelers were scheduled to play the Denver Broncos in a showdown game at Mile High Stadium on November 9, 2009, there was speculation that free safety Ryan Clark might not be in the game.[25] Clark has sickle cell trait. Soon after a game played in Denver on October 21, 2007, Clark became very ill. Some of his blood vessels burst and Clark's spleen was removed as well as his gall bladder. Shortly before the game against the Broncos on November 9, Coach Mike Tomlin deactivated Clark because, despite the importance of the game, he recognized the potential risk to Clark's life.[26]

Tomlin did not deactivate the Steelers wide receiver Santonio Holmes who also has sickle cell trait. Holmes had also played in the October 21, 2007, game at Mile High Stadium.[27] He did not become ill, although, by his own admission, he had real trouble breathing and had an oxygen mask and inhaler right next to him for use when the Pittsburgh defense was on the field. One thing that the Fort Bliss soldiers, Polie Poitier, and Ryan Clark all had in common was that they exercised strenuously at high altitude where oxygen is less abundant. Only Clark survived.

The natural distribution of sickle cell disease generally coincides with the range of malaria caused by *Plasmodium falciparum.* Hence, it is found in Africa; in parts of Greece, Turkey, and the Arabian Peninsula; and in India. Because these different populations have been isolated for a long time, the cause seems likely to have been recurrent mutation in different places. There is evidence that in Africa alone the sickle cell mutation has arisen and been selected independently on four different occasions.[28] A fifth independent mutation has been identified in the Eastern Province of Saudi Arabia that is also found in India.

Malaria and thalassemia

Malaria was once common around the Mediterranean. Herodotus (484–425 BC) and Hippocrates (460–370 BC) both refer to a fever recognizable as malaria.[29] Some Roman writers attributed malaria to foul-smelling fumes given off by the swamps around Rome. In fact, the name malaria derives from the Italian mal'aria and may first have been used in print by Leonardi Bruni in 1476. However, after World War II, DDT spraying programs directed at *Anopheles* mosquitoes successfully eradicated malaria from countries like Cyprus, Greece, Italy, and Spain.

Although the sickle cell gene still occurs at high frequencies in parts of Sicily and southern Italy, and in northern Greece, of much greater concern around the rim of the Mediterranean is another hemoglobin disease called beta thalassemia. This disease was described in 1925 by a tall, aristocratic Detroit pediatrician of austere demeanor and perfect manners named Thomas Benton Cooley and his associate Pearl Lee.[30] Their paper, published in the *Transactions of the American Pediatric Society,* described a disease they had observed in several children of Italian descent as a severe anemia with associated spleen and liver enlargement, discoloration of the skin, and bony changes. Initially, the disease became known as "Cooley's anemia," a name that Cooley frowned upon. Seven years later, George Hoyt Whipple, Dean of Medicine and Dentistry at the University of Rochester, in collaboration with his colleague William Bradford, Professor of Pediatrics, described the pathology of "Cooley's anemia" and gave it a new name "thalassemia."[31] The name derives from "thalassa," Greek for sea and it is clear from the title of a 1936 paper "Mediterranean disease-thalassemia" that they were thinking of the Mediterranean Sea.

Once again, the *HBB* gene is involved as it is in sickle cell disease. It seems likely that the defective *HBB* gene was selected in Mediterranean populations because carriers exhibit at most a mild anemia (thalassemia minor) and are more resistant to malaria than normal individuals. This is borne out by studies carried out in New Guinea where carriers are far more frequent in coastal areas where the incidence of malaria is high, whereas they are very low in the malaria-free highlands.[32] In contrast to the sickle cell gene whose product is a defective β globin molecule caused by a single amino acid change, over 200 mutations are known in the *HBB* gene that can cause β thalassemia.[33]

As is true of sickle cell trait, thalassemia minor usually poses no problem unless a person is subjected to great physical stress. In early September 1996, Pete Sampras, one of the greatest tennis stars of all time, played Alex Corretja of Spain in a grueling quarterfinal match at the U.S. Open. During the fifth set tiebreaker, which Sampras eventually won, he began to wobble unsteadily, gasped for air, and threw up. *Toronto Globe and Mail* reporter Tom Tebbutt guessed correctly that Sampras has thalassemia minor.[34] In fact, Sampras knew he had thalassemia minor, but had said never said anything about it. In his autobiography, he wrote that he would have denied it had Tebbutt interviewed him because he did not want to give his opponents an edge.[35]

Thalassemia major has much more serious consequences than sickle cell anemia because a person with this disease is unable to produce β globin and, therefore, hemoglobin. This is not apparent at birth because, as mentioned earlier, fetal γ globin substitutes for β globin prior to birth, but in the newborn infant, expression of fetal hemoglobin is shut off and β globin begins to replace γ globin. Anemia begins to develop within the first months after birth and a child with this disease becomes dependent on blood transfusions. These must often be given every two to four weeks. However, the repeated transfusions result in an iron overload. A drug called deferoxamine (Desferal) that removes iron from the blood was the standard treatment for many years.[36] Unfortunately, the drug has to be given under the skin using a portable pump hooked up overnight. It is also expensive. The annual cost for an adult in the United States in 2006 was $16,270. In November 2005, a pill, Deferasirox, was approved for use in the United States that serves the same purpose. Sadly, the pill is more than twice as expensive as desferoxamine.

In the Americas, from 0% to 3% of the population are carriers of a defective thalassemia beta globin gene. This increases to 2%–18% in the Eastern Mediterranean.[37] The Island of Cyprus is at the high end as over 14% of the population carry a thalassemia gene.[38] When the island was divided between Greeks and Turks in 1974, Cyprus lacked a blood bank. This meant that parents who had a child with β thalassemia had to find a blood donor who would accompany the child to the hospital once a month. By 1982, a voluntary screening program in the Greek part of the island had cut the number of thalassemic babies born to 8 from 51 in 1974. In 1983, the Greek physicians running the program persuaded the Orthodox Church, which controls weddings on the Greek part of the island, to require a premarital certificate stating that couples had been tested for the thalassemia gene. No thalassemic babies were born in 1986. Only 10 have been born since. One was the result of a misread test and most of the rest were born to parents who declined to have an abortion.

Like the Greek side of the divided island, the Turkish-governed part of the island also introduced mandatory screening programs for β thalassemia. After screening, people on either side can make their own decisions about marriage. If two carriers marry and desire to have children, the woman can voluntarily undergo prenatal screening and request an abortion if the fetus is fated to have the disease. In both the Turkish and Greek regions, the government will pay for the abortion. This policy has the support of both constituencies because of the enormous toll the disease was taking on the population. During the 1970s, both governments agreed to pay the cost of Desferal treatment, but this is no longer true as it would be prohibitively expensive at $1,800 to $5,000 per child per year and more than $7,000 for an adult.

The α thalassemias are a collection of different mutations that result in deficiencies of α globin. As mentioned earlier, there are two genetic loci coding for α globin called *HBA1* and *HBA2*. Thus, any one individual has four copies of these genes: Two are maternally derived and two are paternally derived. Alpha thalassemia usually results from deletion of one or more of these genes. The α^+ thalassemias are quite common, but only produce mild forms of anemia because some α chains are present. In α^0 thalassemia, in contrast, no α polypeptides are produced and fetuses with this condition are stillborn. Like thalassemia minor, the protective effect of α^+ thalassemia against malaria has been demonstrated in both Africa[39] and Papua New Guinea.[40]

Malaria and hemoglobins C and E

Hemoglobins C and E also appear to protect against malaria. Like sickle cell anemia, hemoglobin C results from an amino acid change in position 6 of the β chain of hemoglobin.[41] But in this case, the acidic amino acid (glutamic acid) is replaced by a basic amino acid (lysine) rather than a neutral amino acid (valine) as is true of sickle cell anemia. Hemoglobin C carriers show no anemia and people with two copies of the mutant gene have only a mild anemia that may not be discovered until adulthood. This hemoglobin is restricted to parts of West and North Africa.

Hemoglobin C protects against malaria both in carriers and even more so in individuals with two copies of the gene. However, protection against malaria is achieved by a mechanism completely different from that found with sickle cell anemia. It turns out that normal red blood cells infected with the malaria parasite display a cell surface adhesion protein, PfEMP-1, that is organized in finely distributed knobs. This protein enables infected cells to adhere to the endothelial cells that line the tiny blood vessels of the host circulatory system where the parasite-infected cells are sequestered. PfEMP-1 also causes infected host cells to stick to uninfected red blood cells in a process called rosetting. This produces clumps of cells that can impair blood flow in critical places such as the brain. In infected red blood cells containing hemoglobin C, the knobs are abnormally distributed and less PfEMP-1 protein is displayed. Consequently, these cells adhere poorly to endothelial cells and uninfected red blood cells, meaning they are far less of an impediment to circulation.

Like hemoglobins C and S, hemoglobin E involves a single amino acid change in the β chain of hemoglobin, but this time at position 26 where an acidic amino acid (glutamic acid) replaces a basic amino acid (lysine).[42] Hemoglobin E is abundant in populations ranging from the eastern half of the Indian subcontinent throughout Southeast Asia with carriers sometimes exceeding 60% of the population.

It is also asymptomatic. Published data suggest that the *P. falciparum* may be less likely to invade red blood cells from hemoglobin E carriers than red blood cells from either normal individuals or with two copies of the hemoglobin E gene, but the explanation for this protective effect remains to be determined.

Malaria and glucose-6-phosphate dehydrogenase deficiency

During World War II, primaquine was widely used by soldiers serving in areas where malaria was prevalent.[43] Some soldiers, almost all of them black, could not tolerate the drug. They developed a condition called acute hemolytic anemia, where red blood cells are destroyed faster than they can be replaced. Hemolytic anemia can be triggered by viral or bacterial infections and also by reactive oxygen species. It was later discovered that people who manifested this anemia were deficient in a metabolic enzyme called glucose-6-phosphate dehydrogenase (G6PD) encoded by a gene located on the X chromosome.[44] One of the functions of G6PD is to protect red blood cells against reactive oxygen species. Although the mechanism of action of primaquine is not well understood, it may produce reactive oxygen radicals. This would explain why the drug adversely affects individuals with G6PD deficiency.

G6PD deficiency in humans is the most common genetic condition in the world, affecting an estimated 400 million people. Some 400 variants are known. In 1989, the World Health Organization classified these variants into five groups.[45] Class I variants have severe G6PD deficiency and chronic hemolytic anemia, whereas Class V variants overproduce G6PD. Different populations have distinct mutations, but for our purposes, only two widespread variants need to be considered.[46] *G6PD* A (-) is a Class II variant that reaches frequencies of 15% in sub-Saharan Africa and produces 10%–20% of normal enzyme levels. There is convincing evidence that this variant is associated with 46%–58% reduction in risk of severe malaria for both sexes. *G6PD* Med (Mediterranean) is also a Class II variant, but with less than 10% of normal G6PD activity. As its name implies, it is common around the Mediterranean, presumably because it also alleviated malarial infections before malaria was eliminated in the region after World War II.

Both of these variants appear to be of relatively recent origin. The A- mutation is estimated to have arisen 3–12 thousand years ago while the Med mutation probably arose 1.6–7 thousand years ago. These numbers suggest that they were selected at around the time that *Plasmodium falciparum,* the most deadly malarial parasite, is estimated to have evolved (see the following section).

The older name for G6PD deficiency is favism.[47] This name comes from the fava or broad bean. These beans, cultivated since ancient times in the Old World, produce pods much like peas or lima beans and, as is true of these vegetables, the seeds rather than the pods are eaten. But some people become sick from eating fava beans and even from the pollen of these plants. The reason is that fava beans can induce hemolytic anemia in people with G6PD deficiency because the beans contain the powerful oxidants vicine and covicine.[48]

The recent evolution of the deadliest malaria

Plasmodium falciparum is by far the most deadly of the four species of *Plasmodium* that cause human malaria. It is estimated that *P. falciparum* is responsible for over 500 million cases of malaria each year and as many as one million deaths, mostly of children.[49] A remarkable scientist named Francisco Ayala and his colleagues have now traced the origin of this dread organism. Ayala, a University Professor at the University of California, Irvine, is a former Dominican priest who won the National Medal of Science in 2001.[50]

Ayala and his colleagues have found that *P. falciparum* arose from the chimpanzee species *P. reichenowi.*[51] This event probably occurred about 10,000 years ago at about the time our Neolithic ancestors were leaving hunting and gathering behind in favor of an agricultural existence. This mutational event in the parasite enabled *P. falciparum* merozoites (the infective form of the parasite), in contrast to those of *P. reichenowi,* to infect human erythrocytes. Chimpanzee erythrocytes have two proteins (let's call them A and B, see reference 51 for precise details) that can potentially act as receptors for a surface protein on the *P. reichenowi* merozoites (let's call it R). However, R strongly prefers to bind to A. Human erythrocytes lack A due to a gene mutation, but have B. Hence, for 2 or 3 million years our hominid ancestors may have been protected from *P. reichenowi*. But then in one strain of *P. reichenowi* (let's call it F), a preference for B was selected. This was the strain that gave rise to *P. falciparum*. The time frame for the emergence of *P. falciparum* is also consistent with the appearance of human genetic mechanisms to resist *P. falciparum* malaria such as G6PD deficiency.

HIV resistance: the smallpox connection

We learn from malaria that increasing the frequency of a gene variant that protects against a disease requires a long period of severe and continuous selection. By the same token, once selection is relaxed, it takes a long time for a variant to dilute out of a population even when it is deleterious. So why is it that around 10% of Europeans are resistant to the AIDS virus, which first emerged in Africa a few decades ago?

The explanation seems to relate to smallpox.[52] European populations were subjected to the ravages of smallpox for centuries. The disease had a fatality rate of around 30%. The majority of its victims were children. Because of the persistence of smallpox, one would expect that resistance mutations would have arisen and have been selected for just as they have been in the case of malaria. In contrast, the bubonic plague ravaged Europe a number of times in the past, but it was a sporadic threat that never lasted long enough to select human resistance mutations.

A receptor protein called CCR5 seems to be the key that links HIV-1 and smallpox. This protein is required early in infection for the HIV-1 virus to gain entry into T cells. *CCR5-Δ32*, a deletion mutation, blocks viral infection. This mutation is present at a frequency of around 10% in European populations, but is virtually absent in Africa, Asia, and Middle East, despite the fact that AIDS is rampant in Africa and parts of Asia. Population genetic modeling suggests that the *CCR5-Δ32* mutation arose around 700 years ago and conferred a dominant resistance against smallpox. This link is further strengthened by the fact that the pox viruses gain entry to white blood cells via proteins like CCR5. In short, AIDS resistance seems to be the fortuitous consequence of a residual resistance to smallpox.

Cystic fibrosis (CF) in Caucasian populations

As the great cities of Europe grew up in the eighteenth and nineteenth centuries, the rivers running through many of them served both as sources of drinking water and vehicles for sewage disposal. Sewage was a particular problem for London because the Thames is a tidal river all the way to the British capital, meaning that sewage backed up. The summer of 1858 was the hottest, and probably the smelliest, on record at the time.[53] MPs avoided Parliament because the river stink was so bad. The Prime Minister, Benjamin Disraeli, fled Parliament with a handkerchief over his nose complaining about the "Stygian pool" that the Thames had become.

The presence of raw sewage in rivers like the Thames meant that waterborne bacterial diseases like diarrhea, cholera, and typhoid became scourges. In his book *The Victorians,* A. N. Wilson remarks "the simplest and most life-threatening of all the hazards facing the urban Victorians was the sheer squalor resultant from their failure to understand that cholera, typhoid and typhus fever were waterborne."[54] Later, Wilson mused on the likelihood that Prince Albert, Queen Victoria's consort, succumbed to typhoid fever.[55] London also endured four cholera epidemics between 1831 and 1867.[56]

These considerations caused researchers seeking to determine whether CF confers some beneficial effect on carriers to focus on waterborne bacterial diseases. At first, it appeared that CF carriers might be resistant to cholera toxin using a mouse model in which one copy of the CF gene had been inactivated.[57] It was claimed that carriers had 50% of the normal amount of the CFTR protein and secreted half the amount of fluid and chloride ions in response to cholera toxin. Because diarrhea and dehydration are major consequences of cholera, often leading to death, this finding would clearly indicate an important selective advantage for CF carriers. However, a subsequent study, also using the mouse model, reported normal secretion of fluid and chloride ions in carriers in response to cholera toxin.[58] Furthermore, another study comparing human carriers with normal people found that both groups secreted chloride at the same rate.[59] Hence, resistance cholera toxin appears not to explain why CF carriers might have a selective advantage.

The second waterborne candidate to be nominated for selective agent was *Salmonella typhi*, the bacterium that causes typhoid.[60] Dr. Gerald Pier and a group he led at Harvard Medical School compared the ability of typhoid bacteria to enter untreated cells in tissue culture from a patient with the Δ508 mutation that causes severe cystic fibrosis with identical cells into which the normal CF gene had been introduced. They found that the latter cells took up two to ten times more typhoid bacteria than the cells with the deletion mutation.

Pier and his group then compared normal mice with mice that were carriers of the Δ508 mutation and mice in which both genes were mutant. The carriers took up 86% less typhoid bacteria than the normal mice, whereas the cystic fibrosis mice took up no bacteria at all. These results seem to show rather conclusively that the CFTR protein is required for uptake of typhoid bacteria by epithelial cells and that this process is greatly inhibited in carriers. The results of Pier and his colleagues are supported by another study showing that CFTR is the binding site for typhoid bacteria.[61]

The most recent entry into the field of conjecture concerning the elevated frequency of cystic fibrosis in Caucasian populations focuses on a tuberculosis epidemic that began in sixteenth-century Europe.[62] The lead investigator, Alison Galvani at Yale, was involved in the HIV-smallpox study. Cholera and typhoid were dismissed as failing to exert sufficiently strong selective pressure in favor of tuberculosis. The problem with this argument is that there is no direct scientific evidence that cystic fibrosis carriers tend to be resistant to tuberculosis. There is simply the very tenuous speculation that a deficiency in an enzyme called arylsulfatase B that accompanies cystic fibrosis could be responsible.[63]

So, at this point in time, it is not clear why cystic fibrosis is found at elevated frequencies in Caucasian populations, although it seems highly likely that some sort of selection is involved. The case for typhoid being the selective agent for cystic fibrosis seems the strongest.

French Canadians and the founder effect

In the middle part of the last century, there was a major reawakening of evolutionary thought. It was captured particularly well in four books, *Evolution: The Modern Synthesis* (1942) by Julian Huxley plus another three, all published by Columbia University Press. The first was *Genetics and the Origin of the Species* by Theodosius Dobzhansky (1937); the second was *Systematics and the Origin of Species* (1942) by Ernst Mayr; and the third was *Tempo and Mode in Evolution* (1944) by George Gaylord Simpson. These books synthesized vast amounts of information on the evolutionary process from the fields of genetics, taxonomy, and paleontology.

Mayr's book emphasized the importance of geographic isolation in the origin of species. Mayr was fascinated by ornithology and his early ideas about speciation evolved during collecting expeditions to New Guinea and other parts of Melanesia in the late 1920s and early 1930s. Mayr was also familiar with the work of mathematical population geneticist Sewall Wright, although, not being mathematically inclined, he had not actually read Wright's papers, relying instead on his friend Dobzhansky for a simplified explanation of what Wright was up to.[64]

Wright had proposed that changes in gene frequencies could occur in populations, particular small ones, by a random process he called genetic drift. In his 1942 book, Mayr began to develop a special case genetic drift that he called the founder effect. His point was that the founders of a new population would have only a fraction of the genetic variation of the original population. Furthermore, by chance alone, gene frequencies in a founding population might diverge from those in the parent population (see Chapter 1).

An excellent example of the genetic consequences of the founder effect in human populations can be seen among the French-Canadian inhabitants of the Province of Quebec.[65] The ca. 6 million French-Canadian residents of this province are the descendants of about 8,500 hundred French settlers who arrived in Nouvelle France between 1608 and 1759. The natural increase in this population in relative isolation over four centuries has led to the clustering of at least 22 hereditary diseases in Quebec. The founder effect probably accounts for the high frequency of a number of these genetic diseases, including Tay-Sachs, β-thalassemia, and cystic fibrosis.

Some of these diseases can be followed back to their roots in France. For example, familial hypercholesterolemia is characterized by very high blood cholesterol levels.[66] People with this condition have a high risk of developing coronary artery disease. Mutations in any of four different genes can result in hypercholesterolemia, but mutations in the *LDLR* gene are by far the most frequent. This is also true in French-Canadian populations where a large (15,000 base pair) deletion accounts for 56% of the mutant genes. This deletion has been traced back to the Saumur region in the Loire Valley of France.

The most severe form of hereditary tyrosinemia is prevalent in parts of northeastern Quebec. The symptoms of this lethal recessive disease vary, but they are usually manifested in infancy and can include liver failure and hepatic cirrhosis. This disease results because a gene (*FAH*) encoding one of the enzymes in the pathway leading to the degradation of the amino acid tyrosine is rendered nonfunctional. The disease in Quebec has been traced to a single founder couple.

One normally thinks of Tay-Sachs disease in relation to the Ashkenazi Jewish population, but the frequency of carriers among French Canadians is 5%–7%. The most common form of this terrible disease makes itself evident in infants between three and six months.[67] These children then lose motor skills, vision, and hearing and live only for a few years. The enzyme affected in Tay-Sachs disease is β-hexoseaminidase A. This enzyme breaks down a fatty substance called GM2 ganglioside that belongs to a family of compounds that play a role in cell recognition and communication. It is composed of two dissimilar subunits that are encoded by the *HEXA* and *HEXB* genes.

Tay-Sachs disease arises as the result of mutations in the *HEXA* gene. Although the *HEXA* mutations found in Quebec's Jewish population are characteristic of those found in European populations, the predominant *HEXA* mutation in French Canadians is a 7,500 base pair deletion that may have arisen in a person born in New France.

Although Canada provides universal health insurance, the provincial authorities control health care and the distribution of services. Because inherited genetic diseases are relatively common in Quebec, the government of the province created "The Quebec Network of Genetic Medicine" in 1971 to provide genetic counseling and screening services.

The Acadians

Quebec's French Canadians are only half of the story. Another group of French migrants formed an independent colony in the Maritime Provinces in 1604.[68] These Acadian French had a thriving colony until British colonists took it over in the course of King William's War (1690–1697). As part of the peace settlement, Acadia was returned to France, but during Queen Anne's War (1702–1713), the British regained control of Acadia when it was ceded to Great Britain in the Treaty of Utrecht (1713). Hostilities continued intermittently until the French and Indian War erupted in 1754.

The British burned Acadian houses, accused the Acadians of guerilla action, and demanded that they swear fealty to the British crown or suffer the consequences. Between 6,000 and 7,000 Acadians were expelled from Nova Scotia. Many fled to Louisiana, believing it still to be a French possession. But in 1764, Louisiana was ceded to Spain as part of the settlement at the end of the Seven Years War (1756–1763).

The Acadians or Cajuns, as they now became, brought their own collection of genetic disabilities with them. They included diseases like Friedreich's ataxia, Charcot-Marie-Tooth disease, and, once again, Tay-Sachs.[69] Two different Tay-Sachs mutations have been identified in the Cajun population.[70] The most abundant of these in Cajun populations is identical with the one found in 70% of all Ashkenazi carriers. This mutation has traced back to an ancestral French couple not known to be Jewish who came to Louisiana from France in the early 1700s. The second mutation was brought from Europe about 100 years ago.

Tay-Sachs and other "Jewish" genetic diseases

Tay-Sachs disease is 100 times more frequent in infants of Ashkenazi Jewish ancestry than in non-Jewish infants. Estimates indicate that 1 in 25 Ashkenazi Jews is a carrier and that one in 2,500 infants is born with the disease. The Center for Jewish Genetic Diseases at Mount Sinai Hospital in New York City was founded in 1982.[71] Its mission is (1) to improve the diagnosis, treatment, and counseling of patients and their families having genetic diseases with a high incidence in Jewish populations and (2) to carry out research on these diseases to discover ways of combating them. The Center established the first screening program for couples that may be unsuspecting carriers of Tay-Sachs and several other genetic diseases that are prevalent in the Ashkenazi Jewish population. The Center also provides prenatal diagnosis for at-risk couples. Cities like Chicago[72] and Phoenix[73] also have centers that screen for Jewish genetic diseases.

In 1983, Rabbi Joseph Eckstein, who had the appalling bad luck to have four infants who were born with Tay-Sachs disease, founded an international genetic testing organization called Dor Yeshorim (upright generation in Hebrew) or the Committee for the Prevention of Jewish Genetic Diseases.[74] Its mission is to prevent these diseases from occurring in the Jewish community worldwide.

In its first year, Dor Yeshorim screened 45 people in New York for Tay-Sachs disease. As of 2009, the organization was screening 21,000 people around the world for Tay-Sachs and other genetic diseases that are prevalent in the Ashkenazi community. Individuals are screened at Yeshivas, Seminaries, Bais Yaakovs (elementary and secondary schools for Jewish girls from religious families), and at private screening locations. The screening is anonymous with each individual and his or her results being assigned a number. The participant provides Dor Yeshorim with their date of birth and telephone number. If a couple planning to marry carries the same genetic disease, the match is discouraged. Most carriers never learn their genetic status unless they have a bad match.

Reform and Conservative Jews account for 35% and 26% of the total, respectively.[75] Both forms of Judaism are tolerant of using methods such as abortion following prenatal diagnosis to prevent birth of babies having a genetic disease like Tay-Sachs, and Reform and Conservative Jews choose their marriage partners freely. For Orthodox Jews, who make up around 10% of the American Jewish population, screening is particularly important. This is because of shidduch, the system of matchmaking in which single Jews are introduced to each other in Orthodox communities strictly for the purpose of marriage. Also, Orthodox Judaism generally opposes abortion unless the mother's life is at risk.

An example of where matchmaking can lead is given in a paper written in 2002 for a seminar on Race and Ethnicity by a Columbia University student, Alex Shimo-Barry.[76] She quotes the experience of a man named Israel who spotted a woman across a Brooklyn coat store who looked like an acceptable wife. "She had clean clothes and was very well put together," said Israel. "She wasn't a mess. I thought she was very nice, very appropriate."

Israel had found his potential bride through a matchmaker who told him to find out whether she appeared to be a promising wife by visiting her at her place of work. Israel liked what he saw and told his matchmaker so. Then Israel discovered through a blood test administered by Dor Yeshorim that they were both carriers of the same debilitating genetic disease. He decided he must find a new wife. "I had found out she's very good, very nice mother's girl, everything, all the details. I heard very nice information about her. I was very upset."

Rabbi Eckstein, Dor Yeshorim's founder, is an outspoken defender of the organization's approach.[77] He wrote that by "keeping the results a secret, the testing program avoids the cost of counseling every carrier of a genetic disease. You don't need the counseling, we do the job for you." Dr. Fred Rosner, a professor at Mt. Sinai School of Medicine and an expert on Jewish medical ethics, agrees, "Dor Yeshorim performs a tremendous service." Rosner believes that the aversion of the Orthodox community to abortion makes screening of parents a moral obligation.

But Rabbi Moshe David Tendler, a professor of medical ethics at Yeshiva University, is concerned. "The question arises, when do you stop?" he asks. "There are close to 90 different genes you wouldn't want to have. Will this lead to people showing each other computer printouts of their genetic conditions? We'll never get married."

Tendler is no crank. He is a great advocate of medical progress and an outspoken supporter of embryonic stem cell research and therapeutic (as opposed to reproductive) cloning for experimental purposes. He also has a PhD in biology from Columbia University and teaches Talmudic Law.

One of Rabbi Tendler's major concerns is that a couple whose match is denied will be stigmatized because each of them will be recognized as a carrier of some genetic disease. The rabbi's concerns are borne out by Brooklyn oncologist Mark Levin. "Our personal experience as well as those working within the Orthodox Jewish communities is that (stigma) is a real concern and that stigmatization of carriers for purposes of marriage has occurred." Levin observed that with "high school students, nearly half of carriers felt 'worried or depressed.'" As Christine Rosen has written in the article from which these excerpts were taken, "Stigmatization of Jews based on their genes inevitably raises the specter of eugenics."

Despite these concerns, it is important to point out that Tay-Sachs screening and testing programs have been extremely successful, reducing the incidence of Tay-Sachs disease by 90% wherever they are carried out.[78]

There are at least 18 different genetic diseases that are found at elevated frequencies in Ashkenazi Jewish populations.[79] Although the consensus seems to be that repeated population bottlenecks and random genetic drift accounts for this, one group of four genetic diseases including Tay-Sachs has received a lot of attention because they all result in enzymatic defects in cellular organelles called lysosomes. These spherical structures perform the cellular equivalents of collecting, recycling, and then disposing of cellular garbage far more efficiently than we humans do in our euphemistically named "landfills."

In 1987, Pulitzer Prize winning author and scientist Jared Diamond and Jerome Rotter speculated in a *Nature* article that Tay-Sachs carriers were being selected for tuberculosis resistance under crowded ghetto conditions.[80] Although this hypothesis has since been discarded,[81] Diamond had recognized that there was something odd about the fact that not just Tay-Sachs, but three other genetic diseases of lysosomal enzymes had been selected in Ashkenazi Jews. In 1994, he published another paper in *Nature* titled "Jewish lysosomes." In that paper, he speculated that their might be "selection in Jews for intelligence putatively required to survive recurrent persecution, and also to make a living by commerce, because Jews were barred from the agricultural jobs available to the non-Jewish population."[82] However, Diamond came up with no mechanistic model to explain why carriers of these lysosomal diseases might possess a selective advantage under the right circumstances.

In 2006, Gregory Cochran, Jason Hardy, and Henry Harpending published a paper titled "Natural History of Ashkenazi Intelligence."[83] Harpending is an anthropologist at the University of Utah and a member of the National Academy of Sciences; Cochran is an independent investigator and iconoclast. Cochran et al. formulated a detailed model of improved neuronal growth in carriers of the four Ashkenazi Jewish lysosomal diseases. They argued that additional neuronal growth should foster the high intelligence that characterizes Ashkenazi Jews, providing them with a selective advantage. Cochran and colleagues also extended their model to carriers of four diseases related to DNA repair that are also common in Ashkenazi Jews. Interestingly, the periodical in which they published their paper, the *Journal of Biosocial Science,* had been known as the *Eugenics Review* until 1968.

Naturally, the Cochran paper received quite a bit of press (e.g., *The New York Times, New York* magazine, and the *Economist*) implying as it did a kind of Faustian bargain in which Ashkenazi Jews paid for their brains with the possibility that they might have genetically defective children. Nicholas Wade of the *New York Times* recorded the opinions of two distinguished population geneticists, both being strong proponents of genetic drift.[84] One of them, Neil Risch of the University of California at San Francisco, had published an important paper a couple of years before the Cochran paper came out arguing that the higher frequencies of the lysosomal genetic diseases in Ashkenazi Jews were the result of genetic drift. He had found that when the lysosomal and nonlysosomal genetic diseases were compared using criteria like the number and frequency of disease mutations and the geographic distribution of these mutations, there was no evidence that special selective forces had acted on the lysosomal genetic disease carriers.[85] Dr. Risch was circumspect saying he was not persuaded by the Cochran paper. In an interview with Jennifer Senior of *New York Magazine,* Risch was blunter. "I see no positive impact from this. When the guys at the University of Utah said they'd discovered cold fusion, did *that* have a positive impact?"[86]

In a 2004 paper, Montgomery Slatkin of the University of California at Berkeley published a statistical test for the founder hypothesis and examined its implications for Ashkenazi Jews.[87] Slatkin came to the same conclusion as Risch: namely that the high frequency of the four lysosomal genetic diseases in Ashkenazi Jews could be accounted for by founder effects. Like Risch, Slatkin was cautious in his interview with Wade, but he was much franker when interviewed by Nathaniel Popper of the *Jewish Daily Forward.* "The only controversy among geneticists is how polite to be about this study. I don't know anyone who thinks it's true."[88]

But could it be that there is more to the story? Suppose there is a selective agent, but we don't know what it is. By analogy with the various mutational changes that have occurred in different populations in response to *Plasmodium falciparum* malaria, we might expect that the four lysosomal storage diseases would be found at elevated frequencies in non-Jewish populations living in the same general geographic area where the Ashkenazi Jewish population originated, namely the Middle East.

As it happens, lysosomal storage diseases are also common in Saudi populations except that Tay-Sachs is replaced by Sandhoff's disease (see Table 3–1). Hexoseaminidase A activity is affected in Tay-Sachs disease because the disease inactivates the α subunit of a two-subunit enzyme. Sandhoff's disease inactivates the β subunit of the same enzyme, but, because this also leads to the loss of the α subunit, both Hexosaminidase A and B activities are absent in Sandhoff's disease. Furthermore, these are not the only genetic diseases shared by Ashkenazi Jewish and Saudi Arabian populations. Canavan's disease, familial cholesterolemia, and glycogen storage disease Type I are among other genetic diseases relatively common in both populations (see references in Table 3–1). So, could it be that there is or was millennia ago a subtle form of selection at work? At a minimum, it will be important to compare the so-called Ashkenazi Jewish diseases with those of other peoples inhabiting the Middle East in more detail.

Table 3–1 Common lysosomal genetic diseases found in Ashkenazi Jewish and Saudi Arabian populations

Disease	Ashkenazi Jewish	Saudi Arabian
Gaucher syndrome	+	+
Mucolipodosis IV	+	+
Niemann-Pick disease	+	+
Sandhoff's disease	-	+
Tay-Sachs disease	+	-

Data taken from Ostrer, Harry. "A Genetic Profile of Contemporary Jewish Populations." *Nature Reviews: Genetics* 2 (2001): 891–898. AlAqueel, Aida I. "Common Genetics and Metabolic Diseases in Saudi Arabia." *Middle East Journal of Family Medicine* 6 (2004), number 6. http://www.mejfm.com/journal/Jul2004/Common%20Genetics%20and%20Meta.pdf

In conclusion, genetic disease is not linked to ethnicity. Rather, the so-called "ethnic genetic diseases" are more prevalent in some populations than others either because carriers are or were at a selective advantage or because the frequency of carriers in the population increases by chance when a small group of founders sets out to establish a new community. In many populations where certain genetic diseases are present at elevated frequency, screening programs have been introduced to identify potential carriers and so to prevent the birth of children with serious or lethal genetic defects.

4

Susceptibility genes and risk factors

Lipoprotein particles transport water-insoluble cholesterol and fats through the bloodstream. The particles formed vary in density with the most familiar being high-density lipoprotein (HDL) and low-density lipoprotein (LDL). HDL, also known as "good cholesterol," carries cholesterol to the liver for processing, thus protecting against heart attack. LDL, or "bad cholesterol," circulates in the bloodstream and builds up into plaques in the inner walls of arteries that feed the brain and the heart. These particles are made up of several different lipoproteins called apolipoproteins A through E. A different apolipoprotein (*APO*) gene encodes each of these proteins.

For years, most research on the apolipoproteins and the genes encoding them focused on their role in fat and cholesterol transport. Then, a startling discovery, made by Allen Roses and his group at Duke University, was announced to the world. The enthusiastic title of Natalie Angier's article on August 13, 1993, in the *New York Times* was "Scientists Detect a Genetic Key to Alzheimer's."[1] Roses and his colleagues had been studying 234 people from 42 families and found that those with two copies the *APOE4* variant had eight times the risk of late-onset Alzheimer's than those who had the other two variants of this gene, *E2* and *E3*, or a combination of these two. Even having one copy of the *E4* variant was sufficient to double or triple the risk of late-onset Alzheimer's. Hence, the *APOE4* variant of the gene was a very strong predictor of late-onset Alzheimer's disease.

APOE4 is a relatively straightforward example of a susceptibility gene for a specific disease, but, as we shall see later in this chapter, there are many susceptibility genes for complex conditions like asthma, diabetes, and heart disease. The mechanism by which alterations in a few of these genes cause disease is understood, but for the majority, this is still a black box. Nevertheless, enough has been learned in the last few years about disease genes that it is now possible to classify these genes in a new way. Geneticists have used this information to construct a map of what they have called the "diseasome," the spectrum of diseases and the genes associated with them. We begin this chapter with the story of a susceptibility gene that prevents adults of many races from digesting milk or milk-related products such as cheese properly, but that has been selected against in northern European populations.

Selection against a susceptibility gene: cow's milk and lactose intolerance

Jacqueline Jannota was never fond of milk.[2] When she went off to college in Chicago, she recalls that there "was pizza everywhere, and every time I ate it, I'd be doubled over in pain." Jacqueline's father, a physician, suspected that she was unable to tolerate milk sugar (lactose) because she lacked the lactase enzyme required to process lactose. He therefore suggested that Jacqueline take lactase pills. They worked. She could now eat pizza without ill effects. Thus, Jacqueline Jannota is susceptible to lactose because the gene that encodes lactase has been turned off. As long as she does not eat dairy products or takes lactase pills, this poses no problem.

An estimated 75% of people worldwide are intolerant of lactose as adults. This is because lactase declines rapidly after weaning.[3] However, in northern Europe, the frequency of lactose tolerance is very high. Thus, in Sweden and Denmark, lactose-tolerant individuals make up over 90% of the population. The frequency of lactose tolerance declines from northwest to southeast across Europe. Around 75% of the Caucasian population of the United States is lactose tolerant, reflecting its mixed European origins. Among nonpastoralists such as the Chinese, only around 1% of the population tolerates lactose as adults.

Lactose tolerance results from dominant mutations that prevent lactase production from being turned off.[4] The predominant mutation in European populations seems to be absent in ancient DNA from early Neolithic central Europeans.[5] But with the advent of dairy farming about 7,500 years ago, lactose tolerance began to emerge. Lactase persistence is not restricted to Europeans, but also occurs in certain African pastoralist groups.[6] However, in these cases, the mutational variants are distinct from those seen in Caucasians. Thus, selection for lactose tolerance seems to have evolved independently on several occasions in response to the domestication of cattle and the introduction of cow's milk as a human food.

Is smoking in the genes?

Popular articles on genes and smoking often start out with Mark Twain's famous observation. He said that quitting smoking is easy. He had done it hundreds of times himself. But you might also want to paraphrase St. Augustine's famous supplication to God about chastity. "Lord, make me stop smoking, but not yet."

Smoking is a susceptibility trait for which there is no obvious selective advantage or disadvantage. Because cigarette smoking accounts for at least 30% of all cancer deaths and more than three-quarters (87%) of lung cancer deaths in developed countries, you might question the second part of the previous sentence. However, lung cancer usually appears late enough that most sufferers will have passed through their reproductive years before contracting the disease. This makes the smoking trait selectively neutral in contrast to lactose tolerance, where there is a clear selective advantage in populations that have relied heavily on dairy products for food over the centuries.

Nicotine is both a stimulant and a relaxant.[7] The nicotine from cigarette smoke travels rapidly from the lungs to the brain, spurring the release of the numerous neurotransmitters and hormones that are responsible for its varied effects. Nicotine achieves this stimulus in part by binding to receptors for the important neurotransmitter acetylcholine. So far, 17 different protein subunits have been identified that belong to these receptors. There are two families of subunits designated α and β. They are assembled in different combinations. The *CHRNA* gene family specifies the different α subunits, while the family encoding the β subunits is called *CHRNB*. Obviously, variants in these genes affecting the binding of nicotine either positively or negatively should influence the degree to which nicotine acts as a stimulant and, hence, the extent to which it becomes addictive.

The body also has a system to detoxify nicotine. It is part of the cytochrome P450 (CYP) system.[8] Estimates suggest that around half of all drugs are detoxified by enzymes belonging to this system, most of which are located in the liver. These enzymes also metabolize toxic compounds, including pollutants, solvents, and recreational drugs. The downside to the CYP system is that it converts certain compounds into carcinogens.

The CYP system includes many polymorphic enzymes, meaning they exist in more than one form. One of these enzymes, CYP2A6, converts nicotine to a metabolically inactive compound called cotinine. The rate at which this conversion occurs depends on the particular CYP2A6 polymorphism a person possesses and, hence, the gene variant that person happens to have. Evidently, any gene variant that affects the rate at which nicotine is converted to cotinine is likely to affect smoking behavior.

As discussed earlier (Chapter 2, "How genetic diseases arise"), publication of the sequence of the human genome coupled with the HapMap gave geneticists two powerful new tools beyond the pedigree to hunt for disease genes: the candidate gene approach and genome-wide association studies. In the former approach, the astute geneticist flips through the genomic library to focus on genes likely to be involved in a specific disease or condition. The idea is to compare this region in people with and without the disease or condition to see if there are certain single nucleotide polymorphisms (SNPs) that associate specifically with that disease or condition. The hope is that they will mark genes of interest.

The second approach is totally unbiased and does not focus on the usual suspects. Instead, whole genomes are screened from lots of people with or without the disease or condition in the same way as in the candidate gene approach. Because they are unbiased, genome-wide association studies appeared to hold the promise of turning up disease genes that are completely novel and that would not be found using the candidate gene approach. Both candidate gene studies and, especially, genome-wide association studies are enormously labor intensive. You need only count the number of authors on papers using these methods.

So, what are the genes involved in nicotine susceptibility?[9] Let's begin with the *CHRN* genes and focus an imaginary microscope on the human genome. As we ratchet up the magnification, our attention is attracted to a region on the long (q) arm of chromosome 15. It goes by the slightly awkward name of 15q24.1/15q25.1. We discover that this region contains three *CHRN* genes. They are called *CHRNA3*, *CHRNA5*, and *CHRNB4*. Comparisons between smokers and non-smokers in candidate gene studies and genome-wide association studies of this region have identified several SNPs as being related to nicotine dependence and/or lung cancer (see Table 4–1).[10–12]

Table 4–1 Single nucleotide polymorphisms (SNPs) associated with nicotine dependence, lung cancer, or both in the CHRNA5-CHRNA3-CHRNB4 region of chromosome 15

SNP	Nicotine dependence	Lung cancer*	Gene	Reference**
rs8034191	Not shown	Yes	*AGPHD1*	10
rs16969968	Yes	Not shown	*CHRNA5*	11
rs1051730	Yes	Yes	*CHRNA3*	10,12

*Unlike the other two SNPs, rs16969968 has not been associated directly with lung cancer.
**See commentary in references concerning which studies demonstrate a linkage with nicotine dependence or lung cancer.

An SNP in the *CHRNA5* gene (rs16969968) results in the substitution of the amino acid aspartic acid for asparagine at position 398 in the CHRNA5 protein. This amino acid substitution increases the risk of nicotine dependence. Another SNP in the *CHRNA3* gene (rs1051730) was found to be closely associated with smoking quantity and lung cancer. A variant (rs8034191) in the aminoglycoside phosphotransferase gene (*AGPHD1*), located close to *CHRNA5*, also appears to increase the risk of lung cancer, but so far has not been shown to increase the risk of nicotine dependence.

Although carriers of the SNPs in the *CHRNA3* and *CHRNA5* genes seem to be more susceptible to nicotine dependence than those who lack them, it is important to note that plenty of people who smoke lack the variants. For instance, in one study comparing the association of rs16969968 in European-American and African-American smokers, the frequency of the variant in the former group was 39% and 6% in the latter group.[13] That means that most smokers in both groups, particularly African Americans, lacked the variant.

So, what can we conclude from all of this? It seems that the two variants in the *CHRNA3* and *CHRNA5* genes increase susceptibility to nicotine dependence, but the connection with lung cancer may well be indirect because analysis of four studies of people who have never smoked and have lung cancer revealed no statistically significant association between the 15q24,25 region and lung cancer risk.[14]

What about smoking behavior and the polymorphic *CYP2A6* susceptibility gene? Adolescents with *CYP2A6* variants that slow the metabolism of nicotine become tobacco dependent at relatively low consumption levels.[15] They also smoke less than adolescents whose CYP2A6 enzyme metabolizes nicotine more rapidly. The same is true of adults who metabolize nicotine slowly. Deletion of the *CYP2A6* gene is fairly common in Asia. A Japanese study revealed that people with this deletion smoke less than those with an intact gene.[16] This fits with expectation as, in the absence of the *CYP2A6* gene, they should metabolize nicotine slowly. The corollary is that people with the deletion are also less likely to suffer from lung cancer than those with an intact gene.

Candidate genes for nicotine susceptibility have also been sought elsewhere, for example, in the dopamine and serotonin neurotransmitter pathways.[17] These candidates include the *DRD2* gene that encodes a dopamine receptor. Dopamine receptors are implicated in neurological processes like motivation, pleasure, and cognition, whereas serotonin affects things like mood, sleep, and appetite. For *DRD2*, the results have been equivocal with some studies suggesting an association with smoking and others not.[18] Furthermore, a genome-wide association study found that the region on chromosome 11q23 that includes *DRD2* increased the chance of nicotine dependence, but that the evidence that *DRD2* itself was involved was weak. Instead, a strong association was found with two other genes in the same region that are not well characterized.[19] This shows how genome-wide association studies can often generate surprises that require further explanation.

In view of the ambiguous status of the *DRD2* variant vis-à-vis smoking, it seems that G-Nostics, based in Oxford, UK, may have jumped the gun in testing for smoking-associated variants in the *DRD2* as well as the *CYP2A6* gene. G-Nostics offers a genetic analysis that targets these two genes. It is called NicoTest.[20] However, as the *Times* of London reported, characterization of the universe of gene variants that may be involved in nicotine dependence is still at a fairly early stage. Dr. Saskia Sanderson, a research fellow in health psychology and genetics at University College London, worries about this. "It is almost certain that in 20 years' time we will have tests for 50–100 genes that help to explain why one person smokes and gets hooked and another does not."[21] Dr. Robert West, a professor of tobacco studies at University College, remarks that NicoTest "has been marketed without adequate supportive evidence. It is all very interesting, but not close to being something to consider as a reliable technology."[22] Dr. Mark Tucker, the CEO of G-Nostics, begs to differ. "I find it incredible because nothing is too soon for someone who is trying to quit smoking. Anything that helps them to find a path to quitting can't come soon enough."[23] To be fair, articles cited on the company's Web site from papers like the *Mail on Sunday* and the *Daily Telegraph* were quite laudatory about NicoTest.[24]

According to the G-Nostics Web site, the company has gotten smokers to quit at a rate of about 65% compared with the 35% for the smoking service of the National Health Service.[25] So what accounts for the company's success? It seems likely that the comprehensive online support offered to would-be quitters plays an important role.

Asthma

At the end of November 2005, the death of a 15-year-old girl named Christina Desforges, who lived in Saguenay, Quebec, was widely reported because she was said to have died from her boyfriend's kiss. The supposedly deadly kiss was at first thought to have triggered Christina's peanut allergy because her boyfriend had eaten a peanut butter sandwich hours earlier. Christina's tragic kiss provided a field day for the media. "Allergy Teen's Fatal Kiss" blared the *New York Post.*[26] The *Times* of London's title was bit more staid, "The Kiss of

Death for Girl with Nut Allergy."[27] ABC News headline read, "'Fatal Kiss' Puts Spotlight on Food Allergies."[28]

As it turned out, it was not Christina's peanut allergy that caused her death, but an asthma attack.[29] In March 2006, the coroner, Michel Miron, went public. He said that peanut allergy was not the cause of Christina's death, but he was not yet prepared to give a fuller explanation of why she had died. He told the Associated Press that he felt it necessary to make the statement because the "Canadian Association of Food Allergies intended to use the Desforges case to launch an education campaign. I had to tell them the cause of death was different than first believed." The coroner's report, released in May 2006, concluded that Christina Desforges had suffered a fatal asthma attack that had nothing to do with the supposedly lethal kiss. Furthermore, traces of the active ingredient in marijuana were found in the girl's body, suggesting she may have been smoking pot before she died.

With asthma and allergies, we enter the realm of complex diseases where a variety of genetic risk factors are involved whose contributions are just beginning to be understood. Furthermore, environmental factors also make a decisive contribution. So nature must be assessed in terms of nurture, a phrase coined by the Victorian scientist Francis Galton, who probably lifted it from Shakespeare's play *The Tempest.* In Act IV, scene I, Prospero, the Duke of Milan, is speaking to the spirit Ariel about the man-beast Caliban. "A devil, a born devil, on whose nature nurture can never stick." Fortunately, for reader and writer, if not for the victim, there are usually one or more clear-cut genetic mutations that can cause a complex disease whose mechanism of action can be explained in addition to the many other genetic factors that make contributions. It is these single gene alterations that are stressed in the discussion that follows.

Let's return to asthma.[30] Asthma is a chronic disease that most often starts in childhood. It inflames and narrows the airways in your lungs. Asthma causes shortness of breath, coughing, wheezing, and tightness in the chest. In the United States, over 22 million people suffer from asthma. Symptoms may be mild and disappear on their own or they may intensify so much in an asthma attack that they are lethal as in the case of Christina Desforges. As was true of this young woman, most people with asthma also have allergies.

Factors that determine a person's susceptibility to asthma include genetics, certain childhood respiratory infections, and contact with allergens like those produced by tobacco smoke, dander from cats and dogs, and cockroaches. But dust mites are the main culprits because they possess at least two proteins that can cause allergy and asthma.[31] About 20 million people in the United States are allergic to dust mites. These mites are worldwide in distribution. The American house dust mite (*Dermatophagoides farinae*) is only about 0.5 mm long and, like other dust mites, feeds on flakes of human skin. We produce a lot of dust mite food because the average individual sheds 0.5 to 1 gram of skin per day.

There has been a steady rise in allergic and autoimmune diseases in developed countries over the past three decades.[32] The same is true of asthma. Furthermore, asthma prevalence increases when people move from a rural to an urban environment.[33] Observations like these have been explained by the hygiene hypothesis. This hypothesis proposes that increases in autoimmune diseases and asthma have occurred because we are too clean for our own good. We don't have sufficient exposure to infectious agents, parasites, or just plain dirt any more. If we did, we would be less likely to suffer from allergies and asthma because our immune system would have been kept busy dealing with these external threats.

As science writer Matt Ridley has observed, asthma "is the tip of the iceberg of 'atopy.'"[34] By this, he means that asthma is just one member of the triad of hyperallergic responses that normally begins with eczema and includes hay fever. The allergic response begins when allergen molecules such as dust mite proteins are recognized as foreign invaders by the body.[35] An immune cascade initiates when allergen molecules attach themselves to the mucous membranes of the nose, throat, and lungs. Cells in these membranes "present" the allergen molecules to immune system cells called T cells. The T cells elaborate signaling molecules called cytokines that induce other types of immune system cells, the B cells or B-lymphocytes. These cells then produce immunoglobulin E (IgE), one of the five classes of immunoglobulins found in placental mammals. Within a few weeks after allergen exposure, IgE antibodies specific to the allergen are produced. These IgE antibodies circulate in the bloodstream and some of them attach to cells involved in the inflammatory response called mast cells. When an allergic individual is exposed a second time to the allergen, the IgE bound to the mast cells undergoes a process called cross-linking that causes these cells to release inflammatory chemical mediators like histamine, prostaglandins, and leukotrienes. The inflammatory response ensues and the symptoms of asthma become apparent.

The large number of actors involved in eliciting the immune response suggests that variations in many different genes might affect the process. In fact, a total of 120 different genes have been identified in at least one study as being associated in some way with asthma.[36] Let's consider just three of these as they illustrate how genes having very different functions may be involved in causing asthma. The mechanism of action of the first gene is very well characterized, but less so for the other two reflecting the fact that the products of these genes play many roles within the cell.

Ichthyosis vulgaris is a common inherited, semidominant skin disorder that affects around one in 250 people.[37] People with this disease possess a dry and scaly skin. The latter characteristic accounts for the name because ichthys is Greek for fish. People with this disease frequently suffer from eczema, allergies, and asthma. The reason is that they have loss of function mutations in a gene called *FLG* that encodes an important skin protein, filaggrin.

Human skin consists of two major layers called the dermis and the epidermis. The epidermal outer layer serves as a protective barrier against viruses, bacteria, and fungi, as well as allergenic particles. The epidermis is composed of several layers of cells. At the base are cells called keratinocytes whose major gene products are keratin proteins. The dividing keratinocytes progress through two layers of cells above them and finally reach the surface of the epidermis, where they lose their nuclei and become terminally differentiated corneocytes. These cells pack tightly to form part of the impermeable *stratum corneum.* In the final stages of differentiation, a structure called the cornified envelope forms around each corneocyte. This is integral to barrier function and one of its essential protein components is filaggrin. Filaggrin assists in the aggregation of keratin filaments. This explains the protein's odd name. It is shorthand for *fil*ament *aggr*egation prote*in*. When filaggrin is missing, the keratin filaments do not aggregate properly, the protective barrier becomes defective, and the protective barrier is breached by external allergens that trigger asthma in people with ichthyosis vulgaris.

The *FLG* mutations show how variants in a specific gene can lead to asthma. Two other genes in which there has been recent interest are the vitamin D receptor gene and the *GPR154* gene that encodes a G protein (GPRA).[38] The vitamin D receptor gene (*VDR*) specifies a protein that binds vitamin D. This complex modulates many cellular processes, including important immune system functions. Certain receptor gene variants in the French-Canadian founder population are associated with asthma, but what their role is in causing the affliction is a question for the future.

G protein is short for guanine-nucleotide binding protein. G proteins act like molecular switches. The off switch is a compound called guanosine diphosphate that has two phosphate groups. The addition of a third phosphate group to form guanosine triphosphate turns the switch to the on position. G proteins relay signals from each of more than 1,000 receptors.[39] Mutations in G protein-encoding genes cause a number of inherited disorders. Why the *GPR154* variants are involved in asthma is not known. Working out the precise role of genes like *VDR* and *GPR154*, whose protein products have multiple cellular functions, illustrates the challenges faced by investigators who want to understand the role such genes play in contributing to asthma.

Coronary artery disease

The *Mona Lisa* by Leonardo da Vinci may well be the world's best-known painting. Madonna Lisa Maria de Gherardini, her real name, died in 1516 at the age of 37.[40] The cause of her death was unknown at the time. Leonardo was not only an artist, but was also an inventor and a scientist. His scientific works include a clear-cut description of atherosclerosis. A careful examination of the *Mona Lisa* reveals a yellow spot at the inner end of the left upper eyelid. It is irregular in shape and leatherlike in texture. There is also a soft and bumpy, but well-defined swelling on the upper side of Mona Lisa's right hand beneath her index finger.

Yellowish patches around the eyelids as well as lumps in the tendons of the hands, elbows, knees, and feet are characteristic of an inherited disease called familial hypercholesterolemia (FH).[41] People with this condition develop coronary artery disease much earlier than is true of the general population. Consequently, they may suffer from premature heart attacks. Perhaps Mona Lisa had FH too. If so, she could have died of a heart attack, although death at an early age from a variety of causes was common in her day.

In the 1970s and '80s, the genetic basis underlying FH was worked out by Joseph Goldstein and Michael Brown, who became friends while medical interns at Massachusetts General Hospital in Boston in 1966.[42] They also developed a consuming interest in FH. Why did people with the disease have such high levels of cholesterol in their blood? No wonder they were likely to suffer from heart attacks. Also, FH appeared to be dominant. People with just one copy of the mutant gene suffered from the disease. Children who had two copies of the mutant gene were particularly sad. They had blood cholesterol levels six to eight times higher than normal, almost all of which was in the LDL form. These children often had heart attacks before they reached puberty and generally died of coronary disease before reaching the age of 20.

After being separated for a few years of further training in genetics, Brown and Goldstein both ended up at the University of Texas Health Center in Dallas in 1972. They continued their experiments on FH using cultured cells from normal individuals and a young girl who carried two copies of the defective gene. Brown and Goldstein discovered that both kinds of cells could make cholesterol if bathed in plasma that was low in cholesterol, but, when the surrounding media contained sufficient cholesterol, the normal cells stopped making the compound while the FH cells continued to do so.

Their results led Brown and Goldstein to hypothesize that normal liver cells possess a receptor that recognizes LDL cholesterol and turns off its synthesis when there is sufficient cholesterol around, but this receptor was missing or nonfunctional in FH cells, so they continued to churn out LDL cholesterol. With the help of volunteers, Brown and Goldstein were able to prove their hypothesis for which they were awarded the 1985 Nobel Prize in Physiology or Medicine. Since then, it has been shown that FH usually arises because of mutations in the *LDLR* gene, whose product is the LDL receptor protein. These mutations render the receptor protein incapable of removing LDL proteins from the blood and, thus, of regulating cholesterol levels.

Mutations in several other genes can also result in FH.[43] One of these genes is the *APOB* gene that we met very briefly at the beginning of this chapter. The *APOB* gene encodes both short and long versions of apolipoprotein B known as B-48 (short) and B-100 (long).[44] It is the B-100 form of apolipoprotein B that is of concern to us here. This form of the protein is found in LDLs, as well as in other lipoprotein particles that are involved in cholesterol and lipid transport. At least five mutations are known that affect the B-100 form of apolipoprotein B in such a way that binding of LDLs to the LDL receptor is blocked. As a consequence, cholesterol accumulates to high levels in the bloodstream and coronary artery disease results.

FH affects about 1 in 500 people,[45] but coronary artery disease is the most common form of heart disease and the leading cause of death for men and women in the United States.[46] Hence, despite the unquestioned importance of environmental factors such as diet, it seems likely that a spectrum of genetic variants also affects coronary artery disease, some of which are rare, whereas others are common.[47] There is a coronary-associated disease region on chromosome 9 (9p21), which also shows an association with type 2 diabetes. Hence, it seems likely that there are risk factors common to both diseases. Connections like these form the basis for the diseasome discussed at the end of this chapter.

Diabetes mellitus

Type 1 and type 2 diabetes are the most common forms of this disease.[48] As of the year 2000, a minimum of 171 million people around the world were known to have diabetes. In the United States, type 2 diabetes is by far the most common form of the ailment—accounting for between 90% and 95% of cases.

Type 1 diabetes is an autoimmune disease that results in the destruction of the insulin-producing β-cells found in the islet of Langerhan's in the pancreas. This form of diabetes can affect children or adults, but has been called "juvenile diabetes" because type 1 diabetes accounts for the majority of cases of the disease found in children.

Just under half of the risk for contracting type 1 diabetes can be traced to the most gene-dense region of the mammalian genome, the major histocompatibility complex (MHC) on the short arm of chromosome 6.[49,50] Roughly half of the 149 genes in this region have known functions in the immune system. Almost all autoimmune diseases can be traced to the products of genes encoded in the MHC region. This includes type 1 diabetes and specifically the products of the *HLA* (for human leukocyte antigen) genes of which there are two groups, *HLA I* and *HLA II.* These are highly variable genes encoding proteins of HLA classes I and II, respectively.

Comparison of the *HLA* genes in diabetic and nondiabetic individuals reveals that both *HLA I* and *II* genetic variants are involved in type 1 diabetes. These variants may contribute to the autoimmune destruction of the pancreatic β cells. Molecules of classes I and II present antigen fragments to different subsets of immune system cells. These are respectively called cytotoxic T and T helper cells. If the genetic variants in question recognize β-cell fragments, the immune system will be stimulated to destroy β-cells and type 1 diabetes will ensue. *HLA* genes are not the only type 1 diabetes susceptibility genes. Variants in at least 15 other genes may also contribute to the disease.[51]

Type 2 diabetes is associated with obesity.[52] The disease develops when chronic overconsumption acts in concert with the genetic background to impair the insulin-signaling mechanism.[53] This condition is referred to as insulin resistance. Type 2 diabetes is also accompanied by a deficiency in insulin secretion. This disease is not an autoimmune disease, but it is genetically complex.[54] Gene variants that are involved in type 2 diabetes belong to four different classes. However, the two major classes of variants affect insulin secretion or insulin sensitivity. At least 18 susceptibility genes for type 2 diabetes have been identified, but they account for only 6% of the heritability of this disease (see Table 4–2). Environmental factors also play an important role in type 2 diabetes. Thus, having elevated blood sugar levels, being overweight, and lack of exercise greatly increase a person's risk of developing this disease.

Table 4–2 Estimates of heritability and number of loci for several complex traits

Disease	Number of loci	Proportion of heritability explained
Age-related macular degeneration	5	50%
Crohn's disease	32	20%
Systemic lupus erythematosis	6	15%
Type 2 diabetes	18	6%
HDL cholesterol	7	5%
Height	40	5%
Early onset myocardial infarction	9	3%
Fasting glucose	4	1.5%

Modified from Table 1 in Manolio, Teri A. et al. "Finding the Missing Heritability of Complex Diseases." *Nature* 461 (2009): 747–753.

Type 2 diabetes is an evermore common ailment in the United States.[55] The increase in frequency of the disease is strongly correlated with the increasing proportion of Americans who are obese. Today, there are around 23.6 million Americans who have type 2 diabetes or about 7.8% of the population. Perhaps even more alarming is the fact that 57 million Americans are "prediabetic."

As is true of other complex diseases, unraveling the interplay between the panoply of susceptibility genes associated with type 2 diabetes and the environmental factors that contribute to this disease is going to take time. However, it seems likely that proper attention to environmental factors like obesity can override the risks posed by at least some susceptibility genes for type 2 diabetes.

For every complex disease, there seems to be a spectrum of susceptibility genes, but, as we have seen, there are usually some clear-cut single gene defects that cause the same symptoms, although these generally affect a minority of individuals. One such disease goes by the awkward-sounding name of maturity-onset diabetes of the young or MODY.[56] MODY shows the clear-cut familial pattern of inheritance characteristic of a dominant gene. Symptoms of the disease usually appear during adolescence or in young adults. Mutations in six different genes have been identified that cause MODY. All of them limit the ability of the pancreas to produce insulin.

Are genetic disease variants common or rare?

Two major hypotheses have been proposed to predict the degree to which genes play a role in complex diseases like coronary artery disease or type 2 diabetes.

The first is the Common Disease/Common Variant hypothesis.[57] According to this hypothesis, at each of the major susceptibility loci, there will only be one or a few alleles that predispose an individual to a specific disease. These will be present in high frequency as long as they have no debilitating effect until after people are no longer reproductively active. In fact, such variants may be selectively advantageous during the first few decades of life.

One example of a common variant associated with a disease expressed late in life is the *APOE4* gene discussed at the beginning of this chapter. This genetic variant occurs in about 40% of people who develop late-onset Alzheimer's and is present in about 25% to 30% of the population.[58]

The alternative hypothesis proposes that disease susceptibility results from multiple rare variants in many genes.[59] This is called the genetic heterogeneity or multiple rare variant model. A rather extreme example of a rare variant involves a disease called progeria.[60] Progeria is a genetic disease that is not transmitted, but results from a dominant, spontaneous mutation. A baby born with progeria seems to undergo a greatly accelerated aging process, becoming older looking and more wizened as childhood progresses until death occurs, usually from heart attack, stroke, or atherosclerosis by the age of 13. Hence, death happens before an individual with progeria reaches reproductive maturity.

Most cells in the human body, with the notable exception of red blood cells, possess nuclei. An envelope surrounds the nucleus. Progeria is caused by a dominant mutation in the *LMNA* gene that encodes a protein called lamin that is integral to the nuclear envelope. The mutation creates an unusable form of lamin, meaning that all nuclear envelopes in the body are defective. Precisely how this defect leads to progeria remains to be established.

It seems reasonable to suppose that these two hypotheses need not be mutually exclusive. The frequency of the variant in the population is going to depend on the degree to which it reduces the potential to produce offspring versus the selective advantage it confers either when homozygous or heterozygous. Sickle cell anemia comes to mind once again. Although sickle cell disease very likely reduces reproductive potential, the selective advantage conferred on carriers in malarial regions overcomes this disadvantage. There are undoubtedly far more subtle genetic balancing acts that contribute to the repertoire of susceptibility genes.

Will the real risk factors please stand up?

At the beginning of the twenty-first century, it looked as though genome-wide association studies were going to be the gold standard for the identification of genetic risk factors associated with complex diseases. By identifying different genetic risk factors and attaching a measure of probable risk to each one, a person's genetic risk profile for a complex disease like type 2 diabetes could be spelled out, taking, of course, environmental considerations into account as well. At the end of the decade, after great expense both in terms of money and person power, it seems fair to conclude that this shimmering grail is still out of reach. As Nicholas Wade put it in the *New York Times* in January 2010, although some 2,000 disease-associated SNPs have been identified, "this mountainous labor produced something of a mouse. In each disease, with few exceptions, the SNPs accounted for a small percentage of the risk. A second puzzling feature was that many of the disease-linked SNPs did not occur in the DNA that codes for genes, but rather in the so-called junk regions of the genome. Biologists speculated that these SNPs must play an as-yet-undefined role in deranging regulation of nearby genes."[61] As Wade pointed out, the Common Disease/Common Variant hypothesis assumed that for diseases that usually strike or cause death late in life, such as coronary artery disease or type 2 diabetes, selection does not eliminate "bad genes" because the child-bearing years are over. Hence, to find disease genes, you need only look at the few sites where genetic variations are common, defined as a frequency of at least 1% in the population.

But David Goldstein, of the Duke University Institute for Genomic Science and Policy, and his colleagues argue that natural selection has actually done a good job of getting rid of common disease-producing variants.[62] They hypothesize that the most important disease-producing variants are, in fact, rare, although that cannot always be true as shown by the association of *APOE4* with late-onset Alzheimer's disease.

Goldstein and his colleagues also have done a very important kind of control experiment. They looked for SNPs associated with sickle cell anemia where the mutation responsible for the disease had been identified long ago as falling in the *HBB* gene (refer to Chapter 3, "Ethnicity and genetic disease"). They found that 179 SNPs showed significant association with the disease. These SNPs were spread over a span of 2.5 million base pairs on chromosome 11 in a region that contains "dozens of genes." Furthermore, within this region there are blocks of sequence in which genetic markers remain associated in a nonrandom fashion (see Linkage Disequilibrium in the glossary for more details). Goldstein and colleagues concluded that, "highly significant association signals can travel...to distant genomic regions." This means that genome-wide association studies probably capture a lot of genetic "noise" and that it can be difficult to separate the "signal" from this noise.

A different way of evaluating the results of genome-wide association studies is to ask how much of the heritability of a complex trait can be explained by the genetic markers so far discovered (see Table 4–2). With the exception of age-related macular degeneration where half of the heritability of this common disease is explained by a small number of common variants with large effects, little of the heritability is explained by the loci discovered so far. As we have seen already, in the case of type 2 diabetes, some 18 genetic loci are listed as contributing to the disease, but they account for only 6% of the heritability, implying there is a very long way to go in characterizing the genetic architecture underlying this disease.

On October 9–10, 2010, Matt Ridley titled his *Wall Street Journal* science column "The Failed Promise of Genomics."[63] He wrote that it is true that genomics has made a big difference for rare inherited diseases. "But not for the common ailments we all get. Nor has it explained the diversity of the human condition in things like height, intelligence, and extroversion. If you sign up to have your genes tested today, you will get back a genetic horoscope couched in caveats that would embarrass a fairground fortune teller. 'Slightly above average probability of osteoporosis, etc.'" This a particularly important point for a person to keep in mind who may be contemplating making use of a gene-testing company to find out about their susceptibility to different genetic diseases (see also Chapter 10, "The dawn of personalized medicine"). Unfortunately, as Ridley points out, genome-wide association studies have turned up lots of genes, "but they still explain embarrassingly little."

"I think most people now view genome wide association studies as something we absolutely had to do and have now been done," Goldstein said. "It's fair to say that for many common diseases nothing of very great importance was discovered, but these studies have told us what to do next."[64] That, of course, will be to sequence entire genomes and identify likely mutations in the risk-factor genes themselves or else within regulatory regions outside of these genes. This should soon be possible as the cost of whole genomic sequencing is dropping rapidly (see Chapter 10). However, be cautioned. Although complete sequences will reveal all of the differences between the disease and control populations, the important ones will still have to be sorted out from those that are trivial. In this case, familial studies may help as they should identify alterations common to individuals in the family who have a particular complex disease.

The Diseasome

Nosology, from the Greek word for disease, is the medical discipline that deals with disease classification. Serious attempts to classify diseases began in the eighteenth century with the *Nosologia methodica* of François Bossier de Lacroix.[65] Linnaeus had a go at the problem in his *Genera morborum* as did Erasmus Darwin, the grandfather of Charles Darwin, in his *Zoonomia*. Both of these scientists applied modified versions of Linnaeus's binomial system used for the classification of plants and animals to classify diseases.

Although these classification systems have not withstood the test of time, a workable classification system is essential if one is to make sense of mortality statistics. The statistical study of disease actually began in the seventeenth century with a man named John Graunt who was working on the London Bills of Mortality. These bills were originally drawn up as a means of warning of plague epidemics. What Graunt did was to classify the proportion of children under six who had died by disease. In the nineteenth century, William Farr, a British medical statistician, improved and standardized the disease classification system and promoted its international use. The importance of such a system became clear at the first International Statistical Congress held in Brussels in 1853.

In 1891, the International Statistics Institute, which succeeded the International Statistical Congress, met in Vienna. Its participants appointed a committee, chaired by Jacques Bertillon, Chief of Statistical Services of the City of Paris, to develop a system for classifying the causes of death. Bertillon's system went into general use in Canada, Mexico, and the United States following a meeting of the American Public Health Association in 1898. Subsequently, in August 1900, delegates from 26 countries arrived in Paris for an international conference whose purpose was to revise the Bertillon system.

So far, the classification system had been directed strictly at mortality, but earlier on William Farr had recognized the desirability of extending the statistical classification system to diseases that, though not fatal, caused disability. This would greatly broaden the classification system to include the kinds of diseases that spread in places like armies, hospitals, and public institutions. A classification system for mental diseases affecting the inmates of lunatic asylums also seemed a good idea. At the 1900 International Conference, the need for a uniform system for classifying diseases as well as mortality was first recognized.

Over the years, the classification systems for diseases and mortality have been revised further. In 1946, the World Health Organization was charged with the responsibility of setting up a worldwide system of classifying the causes of disease and mortality. The result was the *International Classification of Diseases, Injuries, and Causes of Death*. Since 1946, this classification has undergone ten revisions with the last being in 1990.

But as our knowledge of the effects of human molecular genetics has increased by leaps and bounds, an entirely new way of looking at diseases has evolved. This has crystallized into the human disease network or diseasome. The idea is that one can connect different human diseases into a network by the genes that affect them.[66]

The diseasome map can be explored online courtesy of the *New York Times* at www.nytimes.com/interactive/2008/05/05/science/20080506_DISEASE.html.

The diseasome opens a new way of thinking about genetic disease, particularly complex diseases, some of which overlap each other, such as type 2 diabetes and coronary disease. There is no doubt that we are going to learn a lot more about the interactions of genes and disease as the diseasome becomes refined in evermore increasing detail. Exactly how the diseasome will be reconciled with the current system for classifying diseases remains to be seen, but it could give the system an entirely new dimension.

5

Genes and cancer

In his State of the Union address in January 1971, President Richard Nixon announced his intention to request an appropriation of $100 million "to launch an intensive campaign to find a cure for cancer, and I will ask later for whatever additional funds can effectively be used. The time has come in America when the same kind of concentrated effort that split the atom and took man to the moon should be turned toward conquering this dread disease. Let us make a total national commitment to achieve this goal."[1]

On December 23, 1971, President Nixon signed the National Cancer Act into law, declaring, "I hope in the years ahead we will look back on this action today as the most significant action taken during my administration." The National Cancer Act, "The War on Cancer," gave the National Cancer Institute a unique autonomy within the National Institutes of Health to deal with the many manifestations of the disease that it has managed with skill and great success.

Usually when we think of war, we think of a defined enemy, for example, the Axis nations in World War II. But the War on Cancer is more like the War on Terror. Both phrases conjure up dread images. On the one hand, we think of turbans, beards, and car bombs, but on the other hand, we think of uncontrolled cell growth and metastasis. But both these phrases oversimplify enormously complex problems. We are only beginning to understand the array of tribal societies that inhabit places like Afghanistan, Pakistan, Somalia, and Yemen with their strange mores, odd customs, and difficult languages. Similarly, cancer is a large collection of different genetic diseases, each of which has its own unique characteristics. Nevertheless, we are gradually coming to grips with the many forms of this most protean of diseases.

This has only been possible because our basic understanding of the molecular and genetic basis of cell function, interaction, and the control of cell proliferation has advanced so much in the past few decades.

This chapter begins with a discussion of skin cancer. Skin cancer will be familiar to many readers of this book who, like the author, must make regular trips to their watchful dermatologist. The genetics of cancer is far too massive a subject to be covered in any detail in a book of this kind so the rest of this chapter is devoted to several different types of cancer. They are chosen, with one exception, to illustrate different aspects of the biology of this complex disease.

Cancer and the skin

Skin is an epithelial tissue. Epithelial tissues are the protective layers of cells that line your inside (e.g., the intestine) and outside (e.g., skin). Depending on their function, epithelial tissues can derive from any one of the three main primitive cell layers of the embryo. Thus, skin arises from the outermost layer, the ectoderm; the ovaries derive from the middle layer, the mesoderm; and the epithelial layers of internal organs like the lungs, liver, pancreas, and gall bladder as well as the lining of the stomach and intestines derive from the innermost cell layer, the endoderm.

Actinic keratosis is by far the most common lesion with the potential to become malignant.[2] An actinic keratosis is usually a small (2–6 mm), reddish, rough or scaly spot of thickened growth (keratosis) caused by significant exposure to sunlight (actinic). Actinic keratoses can either regress or remain the same, but occasionally develop into squamous cell carcinomas. Dermatologists frequently prescribe Fudex to treat actinic keratoses. This is actually a compound called 5-fluorouracil. It interferes with the action of an enzyme called thymidylate synthetase that is required for the synthesis of thymine-containing nucleotides and, hence, for the replication of DNA. As a consequence, the dividing cells in the actinic keratosis are killed because they make defective DNA molecules.

Squamous cell carcinomas are the second most common form of skin cancer. They normally do not cause problems if caught and treated early. Squamous cell carcinomas can also arise in other kinds of epithelial cells in addition to skin cells. For example, most cases of head and neck cancer result from squamous cell carcinomas that initiate in the moist, mucosal cells that cover the surfaces of the mouth, nose, throat, and esophagus.[3] Head and neck cancers account for around 3% to 5% of all cancers in the United States. Adenocarcinomas are the other big class of carcinomas. They arise in specialized cells that have a secretory function that line internal organs. For example, adenocarcinomas arise in the ducts and lobules of the breast. They are glandular tissues that make breast milk. In fact, carcinomas are the most common of all human cancers, accounting for 80% of cancer deaths in the West.[4]

Basal cell carcinoma is the most common type of skin cancer.[5] Unlike squamous cell carcinomas for which actinic keratoses are the precursors, there are no clear precursors for basal cell carcinomas. UV radiation is the ultimate culprit for all three of these conditions. Most of these cancers are spontaneous (sporadic) in nature, but some are hereditary. Over 90% of squamous cell carcinomas and 50% of basal cell carcinomas have mutations in the *TP53* tumor suppressor gene that encodes the p53 protein.

Tumor suppressors are one of three classes of genes involved in cancer. The other two classes are called oncogenes and repair genes. Oncogenes are derived by mutation from proto-oncogenes. These are normal genes that help regulate cell growth and differentiation. Oncogenes have been compared with stuck gas pedals in that their protein products accelerate cell division and growth in an uncontrollable fashion. Tumor suppressor proteins act like car brakes. They keep cell division under control and slow it down in response to perceived defects arising in DNA during replication or in the apparatus that ensures proper chromosome distribution during mitosis. Repair genes are like mechanics. The proteins encoded by them fix defects that arise in the cellular DNA, in the apparatus required to maintain accurate distribution of the chromosomes during mitosis, and so on.

The p53 protein has been portrayed as the guardian of the genome. Many forms of cellular stress can activate the protein. For instance, if damage to the DNA has occurred and been recognized, p53 can arrest cell growth at the point in the cell cycle prior to DNA synthesis in preparation for cell division. The p53 protein can also activate DNA repair proteins to fix the damage. And it can initiate the process of apoptosis or programmed cell death if the damage cannot be fixed.

Once the guardian is inactivated, as the result of a *TP53* mutation in a basal cell or squamous cell carcinoma, there is no way to close the gate (i.e., arrest the cell cycle). This permits chromosomal damage to be passed on to daughter cells as long as it is not lethal to them. This is the first step in genomic destabilization. It is followed by chromosomal rearrangements and changes in chromosome number. New mutations that may affect oncogenes and other tumor suppressors also go uncorrected. For example, mutations in another tumor suppressor gene called patched (*PTCH*) play a major role in basal cell carcinoma formation.[6] The *PTCH* gene product is a receptor protein that spans the cell membrane (see reference 6 for a more detailed description of what happens when *PTCH* mutates).

Although the great majority of cancers are sporadic, a familial cancer can usually be identified that relates to the broken gene. For example, loss of p53 function is one characteristic of over 50% of all cancers.[7] The familial cancer associated with this loss of function is called Li-Fraumeni syndrome.[8] The disease shows a dominant pattern of inheritance and greatly increases the risk of developing several types of cancer, particularly in children and young adults. These commonly include breast cancer, osteosarcoma, a form of bone cancer, and cancers of soft tissues like muscle. Cancers of the latter type are referred to as sarcomas as opposed to the carcinomas so far under discussion. Sarcomas amount to only about 1% of all cancers and share a common origin in the mesoderm.[9] Sarcomas affect tendons, muscle, bone, and other connective tissues.

Let's consider for a moment what dominance means in the case of Li-Fraumeni syndrome. Dominance is pretty clear-cut in the cases of diseases like cystic fibrosis or Tay-Sachs. The normal form of the gene is dominant to the disease gene. But dominance in the case of familial cancers usually means something else. It means that a person is carrying a damaged form of the cancer gene, in this case *TP53*, so there is only one instead of two good copies of the gene. Suppose the probability of a mutation in the *TP53* gene is 1 in a million (1 in 10^6) per cell generation. The probability of a cell containing two mutant

TP53 gene mutations is the product of the individual probabilities or one in a trillion (1 in 10^{12}). On the other hand, if one copy of the gene is already mutated, the probability of a mutation inactivating the good gene is one in a million. That might seem like a big number, but it is not if you consider the large number of epithelial cells in the body and the fact that over a lifetime they divide a great many times.

The familial disease associated with *PTCH* mutations is called Gorlin syndrome.[10] Compared with the number of people who develop sporadic basal cell carcinomas, only 1% suffer from Gorlin syndrome. Individuals with this disease begin to develop basal cell carcinomas during adolescence or as young adults. The number of basal cell carcinomas varies among affected individuals. Generally speaking, people with lighter skin develop more basal cell carcinomas than those who are darker in color. People with Gorlin syndrome often develop benign tumors of the jaw and are also at higher risk of developing other tumors than the general population. The high incidence of basal cell carcinomas in Gorlin syndrome is exactly what one would expect based on the observation that *PTCH* is often mutated in sporadic basal cell carcinomas.

In the early 1970s, Alfred Knudson, a pediatrician who was interested in childhood cancers, first made the distinction between sporadic and familial cancers.[11] Knudson and his colleague Louise Strong were studying a childhood eye cancer called retinoblastoma. Prior to 1866, this disease was lethal as the cancer spread to the brain and most children with the disease died before the age of ten. However, with the invention of the opthalmascope in that year it became possible for doctors to peer into the eye and detect the incipient mass of cancer cells at the back of the eyeball. The disease could now be prevented from spreading, but the price was very high as it involved surgical removal of the eyes. If these blinded children went on to have families of their own, it turned out that some of their children also had retinoblastoma.

By the time Knudson started his own work on retinoblastoma, ophthalmologists were able to kill the cancerous cells with radiation if the tumor was caught early enough. And now, Knudson came up with a model for what was happening. It was called the "two-hit" hypothesis. The first "hit" was the mutant gene that was familial in transmission, but a tumor did not begin to develop until a retinal cell had received a second mutational "hit" that disrupted the normal copy of the gene. In confirmation of this model, Knudson and Strong found that, in malignant retinal cells from patients with the disease, both copies of chromosome 13 were lacking the same piece of DNA. Today we know that the key gene that is missing is the *RB* gene, another tumor suppressor.

Let's return to squamous cell carcinomas one last time. Amplification of a gene called *MYC* has been noted in some actinic keratoses, but more frequently in squamous cell carcinomas (for more explanation on *MYC*'s oncogenic properties, see reference 12). When *MYC* is either deregulated or amplified, it behaves like an oncogene. The gene is frequently amplified in squamous cell carcinomas, but it is deregulated in 15% to 30% of all other human cancers.[13]

The most serious form of skin cancer is malignant melanoma. Most Australians (Aborigines excluded) arrived in Australia from England in the eighteenth and nineteenth centuries having been "transported" as criminals or other sorts of riffraff. These pale-skinned people and their descendants were very sensitive to the intense Australian sun and, hence, very susceptible to skin cancer. Because moles and melanomas can be confused, the Australasian College of Dermatologists has formulated easy-to-follow, but detailed guidelines for moles and melanoma.[14] The College points out that many Australians have up to 100 moles that are 2 mm or more in size by the age of 15 years. People with greater than 100 moles are at increased risk of developing melanoma. This does not mean that melanomas develop from the moles themselves, although they sometimes do. Many melanomas arise as new spots on the skin.

Approximately 90% of melanomas are sporadic,[15] but 10% of melanoma cases are inherited. At least 25% of hereditary melanomas can be attributed to mutations in a tumor suppressor gene called *CDKN2A*. The *CDKN2A* gene can yield two different protein products depending upon how it is "read." One of these, called p16, inhibits enzymes called cyclin-dependent kinases that are involved in cell cycle regulation. The second protein product of *CDKN2A* is called p14ARF. This protein acts to stabilize p53 by preventing degradation of this protein by a protein called MDM2.[16] Hence, a mutation that inactivates both *CDKN2A* functions will affect regulation of the cell cycle while, at the same time, hastening p53 degradation and, therefore, preventing the protein from exercising its guardian function.

In other melanomas, two oncogenes are responsible for cell proliferation. The products of these genes, called *NRAS* and *BRAF*, act primarily through a protein called RAF (see Table 5–1). RAF is the first protein in a cascade that leads to increased gene transcription and cell growth. *NRAS* gene-activating mutations occur in 10% to 20% of melanomas, whereas mutations that activate the *BRAF* gene are found in 50% of melanomas. *NRAS*, but not *BRAF*, also activates the P13K-Akt pathway, stimulating growth. The ability of P13 to activate Akt is blocked by the product of the *PTEN* tumor suppressor gene. *NRAS* mutations can activate this pathway, but in 30%–50% of melanomas, this pathway is activated by deletion or mutation of the *PTEN* gene. Lastly, *AKT3* amplification results in Akt activation in roughly 60% of melanomas. What these overlapping percentages mean is that melanomas usually have multiple genetic alterations—any one of which can lead to the disease. Hence, attempting to target a specific gene defect is unlikely to be effective as a therapy.

Table 5–1 Some gene alterations commonly found in melanomas

Gene	Type	Genetic change	% of melanomas
NRAS	Oncogene	Mutation	10–20
BRAF	Oncogene	Mutation	50
PTEN	Tumor suppressor	Mutation or deletion	30–50
AKT3	Oncogene	Amplification	60

Data from Chudnovsky, Yakov, Paul A. Khavari, and Amy E. Adams. "Melanoma Genetics and the Development of Rational Therapeutics." *The Journal of Clinical Investigation* 115 (2005): 813–824.

Melanoma of the skin varies markedly in frequency by race. Thus, the incidence of the disease per 100,000 among white males is 28.9 compared with 1.1 for African Americans and 4.6 for Hispanics. Yet it seems highly unlikely that melanoma of the skin has acted as a selective agent for increased skin pigmentation as the median age for diagnosis is 59 years of age, well past the years of maximum reproduction.[17] So why is dark skin favored in regions of the world where the sun is at its brightest?

It seems likely that our early hominid ancestors had black fur and light skin like the chimpanzee.[18] As our African ancestors evolved further, they became bipedal and hairless and acquired a modern mechanism for sweating. At the same time, their skin acquired more and more UV-absorbing melanin pigment. The effects of UVB radiation are especially harmful (refer to Chapter 2, "How genetic diseases arise"). Not only does irradiation in this wavelength range cause sunburn, but UVB-induced damage to sweat glands disrupts sweating and, thereby, thermoregulation. Yet even this might not be enough to explain the selection of dark pigmentation in our ancestors.

Another explanation involves the essential nutrient folic acid that is converted to folate in the body. Folate is essential for nucleotide synthesis and, therefore, the replication and repair of DNA. Folate deficiency has been linked to fetal malformation in nonhuman mammals and to neural tube defects in humans. These can lead to anencephaly, a lethal condition by the time of birth, craniorachischisis, a condition causing embryonic development to fail early on, and spina bifida. A folic acid supplement prevents 70% of these neural tube defects. Folate has also been shown to be critical for spermatogenesis in mice and rats. Because folate is exquisitely sensitive to UV light, it may be that dark skin pigmentation evolved to prevent folate degradation in response to reduced rates of embryonic survival and spermatogenesis.

In contrast, exposure of the skin to the UV rays in sunlight causes the conversion of 7-dehydrocholesterol to previtamin D and subsequently to vitamin D. Vitamin D deficiency is the primary cause of rickets, a very common childhood disease in developing countries. Lighter skin pigmentation may have been selected at higher latitudes where light intensities are lower in response to vitamin D deficiency.

Six or more genes are involved in controlling skin color. Genetic investigations suggest that light skin color evolved independently in European and East Asian populations.[19] This finding implies that, even if the vitamin D hypothesis is wrong, there is selection for light skin color in the northern regions of the earth.

Finally, there is no better way to illustrate the role of repair enzymes in preventing cancer than to consider a group of rare autosomal recessive diseases referred to collectively as xeroderma pigmentosum.[20] Individuals with xeroderma pigmentosum are at a thousandfold greater risk of UV-induced skin cancer before the age of 20 than normal individuals. People with this disease have to stay out of sunlight. Mutations that cause xeroderma pigmentosum do not affect the gene encoding the photolyase enzyme. This enzyme requires visible light to repair damage to DNA (see Chapter 2). Instead, they principally affect repair genes in one of the "dark repair" pathways (i.e., light is not required for repair) called the nucleotide excision repair pathway. The enzymes coded by these genes remove damaged nucleotides and facilitate insertion of undamaged nucleotides. People with xeroderma pigmentosum lack a functioning excision repair pathway, so they are unable to repair UV-damaged nucleotides; mutations can, therefore, occur in oncogenes and tumor suppressors, posing a serious risk of skin cancer.

The virus connection: cervical cancer

All women are at risk for cervical cancer, but the disease is uncommon in women under 30.[21] In the United States, the median age at diagnosis is 47. In 2006, 11,982 women in this country were diagnosed with cervical cancer and 3,976 died from the disease. The cervical cancer rate varies somewhat with race and ethnicity with the incidence for all races being 8.0 per 100,000. Worldwide, 500,000 women contract cervical cancer each year with 80% of cases being reported in the developing world.[22]

Cervical cancer is caused by human papillomaviruses (HPV). There are more than 100 kinds of HPV.[23] Certain types of HPV cause warts, or papillomas, hence the name. These are benign tumors. Genital HPV infections are sexually transmitted and quite common. Over 30 types of HPV are transmitted sexually. Two of these, HPV-6 and HPV-11, are frequently associated with genital warts. With respect to cervical cancer, HPV strains separate into "low-risk" and "high-risk" groups. That latter group includes at least 15 different types, of which HPV-16 and HPV-18 are responsible for about 70% of cervical cancers.

In 1928, a Greek pathologist named George Papanicolaou reported that he could tell the difference between normal cells and cancerous uterine cells by means of a simple, vaginal smear.[24] Papanicolaou had emigrated from Greece to the United States in 1913, accepting a position in the Department of Anatomy in Cornell Medical School in New York City. In 1943, Papanicolaou published a monograph titled "Diagnosis of Uterine Cancer by the Vaginal Smear" in collaboration with Dr. Herbert Traut of the Department of Obstetrics and Gynecology at Cornell Medical School. This was the paper that led the medical community to recognize the importance of the discovery that Papanicolaou had made much earlier. Papanicolaou's diagnostic procedure, since named the Pap test or Pap smear, has been the standard procedure in gynecological examinations for detection of cervical cancer for many years now. If caught early, cervical cancer is readily treated surgically.

The realization that there is a relationship between viruses and cancer dates back to 1909 when Peyton Rous, working at the Rockefeller Institute in New York City, began the research that led to the discovery of the first tumor virus.[25] Rous was studying a sarcoma that had appeared in the breast muscle of chicken. He found that if he ground up the cancerous tissue, filtered it, and injected it into young chickens, they developed tumors too. Because the infective agent could be passed through a fine filter, it could be no bigger than a virus. Rous published his results in 1911, but had to wait 55 years for the ultimate recognition of the importance of his discovery when he received the Nobel Prize in Physiology or Medicine in 1966. The virus was by now known as the Rous sarcoma virus (RSV).

In the early 1950s, Harry Rubin in Renato Dulbecco's laboratory at the California Institute of Technology discovered that, when cultured embryonic fibroblast cells from chicken were infected with RSV particles, they did not kill their host cells like most viruses, but continued to produce virus particles over a period of days, weeks, and even months. Unlike normal cells in tissue culture that cease to grow after they form a confluent layer of cells over the surface on which they were inoculated (contact inhibition), these infected cells continued to grow and pile up on one another (transformation).

A graduate student of Dulbecco's, Howard Temin, continued to work on RSV together with Rubin. Their results showed that the transformed cells retained the virus, but this did not necessarily mean that the virus had transformed the cells it had infected. An experiment done at Berkeley involving a mutant of RSV that was sensitive to a temperature of 41°C, but grew normally at 37°C settled the question. As long as the cells were grown at 37°C, they retained their transformed character, but when they were switched to 41°C, they reverted to the shape and growth pattern they had possessed before being infected with RSV.

RSV particles were known to contain single-stranded RNA, rather than double-stranded DNA, as their genetic material. How was the virus maintained in the transformed cells in this state? In 1960, Temin began to formulate the provirus hypothesis, which he formally proposed in 1963.[26] Temin's hypothesis seemed like heresy because he had postulated that the RNA of RSV was copied into a DNA provirus, which could in turn produce RNA copies. At the time, the central dogma of molecular biology held that the information in DNA was transcribed into messenger RNA whose information was then translated into protein. An exception had to be made for RNA viruses. It was assumed that these viruses transferred their information from RNA to RNA and from RNA to protein, but a reversal of information transfer from RNA to DNA was a bridge too far. Temin's hypothesis was ignored for over six years. Then in 1970, both he and David Baltimore showed independently that RSV RNA encoded an enzyme called reverse transcriptase that could make a DNA copy from the viral RNA. Baltimore, Temin, and Dulbecco shared the Nobel Prize in Physiology or Medicine in 1975.

Within a year of the discovery of RSV reverse transcriptase, proviral DNA was detected within the chromosomal DNA of RSV-infected cells. But a piece of the puzzle was still missing. Why did the presence of the RSV genome turn normal cells into cancer cells? J. Michael Bishop and Harold Varmus solved this problem in their joint laboratory at the University of California, San Francisco. One of the viral genes must encode a protein that transformed normal cells to cancerous cells. The hypothetical gene was named *src*, short for sarcoma. A comparison of the RSV genome with other retroviral genomes that did not cause cellular transformation revealed that there was only a little information left in the RSV genome to encode the *src* gene, as most of the genome was devoted to making necessary viral proteins.

Bishop and Varmus made a DNA probe for this small region and, much to their surprise, they found this DNA sequence not only in cells infected by the virus, but in uninfected cells too. In other words, the virus had hijacked a cellular gene. It turned out that the cellular gene (c-src) was part of the normally functioning genome of the cell, a proto-oncogene, but when captured by the virus (v-src), it became an oncogene. The product of this gene was sufficient to transform a normal cell into a cancer cell. In 1989, Harold Varmus and J. Michael Bishop received the Nobel Prize in Physiology or Medicine for this work.

It is time now to return to cervical cancer. A great deal of work was also being done on DNA viruses suspected of producing cancer. These included a virus that caused papillomas (warts) on the skins of rabbits discovered by Richard Shope and named for him, a mouse virus called polyoma, and a monkey virus with the name of SV (for simian virus) 40. Although these viruses usually followed a normal life cycle in which they reproduced in an infected cell and ultimately lysed the cell, there were certain kinds of cells on which they could not grow. However, occasionally, one of these cells out of thousands would be transformed in the same way that RSV transformed cells. Like the DNA copies of RSV, the DNA of these viruses also became associated with the chromosomal DNA of the cell. This group of tumor viruses or family originally received the name papovavirus for *pap*iloma, *po*lyoma, and the *va*cuoles caused by SV40 virus when it went through its productive cycle that resulted in cell lysis. Today, these viruses are separated into the viral families Papillomaviridae and Polyomaviridae.[27]

Human papillomaviruses (HPV) are small, circular DNA viruses.[28] The HPV genome consists of nine overlapping genes and a control region. The control region and the seven genes (E1–E7) expressed early on following viral infection (early region) are involved in the regulation of the viral life cycle and in the transcription of viral DNA. The other two genes (L1 and L2) are expressed subsequently (late region). These genes specify the major and minor proteins of the coat (capsid) of the virus that encloses the viral DNA.

HPV virus particles (virions) initially infect the basal cells of the epithelium. The viral DNA is maintained independently of the host genome in these infected cells at a copy number of about 50–100 viral genomes per cell. As infected daughter cells migrate to the upper layers of the epithelium, the late genes are expressed, new virus particles are formed, following which the progeny virus are released and able to infect new cells. The role of the E6 and E7 proteins is essentially to prevent the immune system from eliminating the virus by blocking interferon regulatory proteins from activating interferon proteins and, thus, stimulating the immune response.

Although a woman's lifetime risk of cervical HPV infection is 80%, most of these infections are resolved spontaneously by the immune system. Integration of the viral genome into the host genome is a rare event, but once it happens, transformation of the infected tissue into cervical cancer proceeds rapidly. In this case, the role of the E6 and E7 proteins is very different. E6 promotes the degradation of p53, whereas E7 disrupts the function of the retinoblastoma protein so the two viral proteins contribute to cancer formation by interfering with cell cycle control. Thus, cervical cancer is caused by two proteins encoded by the viral genome in contrast to RSV, which induces cancer via a hijacked host protein.

If a woman's pap smear indicates cervical cancer, a number of treatment options are available. They include surgery, liquid nitrogen freezing of the tumor, and carbon dioxide laser destruction (ablation). The most promising preventive method at present involves vaccination against HPV types that are highly likely to cause cervical cancer.

In 2010, freelance science writer Rebecca Skloot published a book titled *The Immortal Life of Henrietta Lacks*.[29] The subject of her book was a young black woman named Henrietta Lacks who had died from cervical cancer. While Lacks was being treated at Johns Hopkins University hospital, a young doctor, George Gey, snipped samples from her cancer without telling her. Gey, who was head of tissue culture research at Hopkins, had been trying for years to culture an immortal line of human cells. Gey found that these cancer cells from Henrietta Lacks, which he called HeLa cells, could be cultured indefinitely. Later, Henrietta's cancer cells would play a crucial role in the development of a polio vaccine; they traveled on the first space mission to see what would happen to human cells exposed to zero gravity; and they would find many more important uses as well. Of course, poor Henrietta Lacks never knew what uses her "immortal" cells had been put to and neither did her family. But the family finally did learn when Rebecca Skloot told the story of Henrietta's cells to her family while she was in the process of interviewing them.

Breast and ovarian cancer and the problem of gene patents

In 2009, the American Cancer Society estimated that 192,370 new cases of invasive breast cancer would be diagnosed in the United States and that an additional 62,280 breast cancer cases would be detected before the disease spread.[30] In that same year, 40,170 women were expected to die from breast cancer with only lung cancer killing more women.

Although a number of genes have been implicated in causing breast cancer, two in particular stand out as villains: *BRCA1* and *BRCA2*. An estimated 5%–10% of breast cancers are hereditary with the rest being sporadic, with *BRCA1* and *BRCA2* being the genes most often implicated in hereditary breast cancer.[31] These genes are usually referred to as being dominant for breast cancer, but in fact they are dominant only in the sense we used the word earlier in this chapter that a person with a breast cancer mutation in one copy of, say, the *BRCA1* gene is quite likely to suffer from breast cancer because a second mutation strikes the good copy of the gene. The protein products of both the *BRCA1* and *BRCA2* genes are involved in repairing double-stranded breaks in DNA, although *BRCA1* is involved in a number of other important cellular activities as well. The probability that a woman will suffer from breast cancer if she carries a mutation in either of these genes increases dramatically with age (see Table 5–2).

Ethnicity plays a role in hereditary breast cancer. Thus, 1% of Ashkenazi Jews have a specific deletion mutation in the *BRCA1* gene. Ashkenazi Jews also have deletion mutation in the *BRCA2* gene at a frequency of 1%. Icelandic women have a different *BRCA2* deletion mutation at the rather high frequency of 7.7%.

The fact that *BRCA1* and *BRCA2* mutations play such an important role in hereditary breast cancer means that testing for mutations in these genes assumes great importance, particularly in at-risk groups. When Genae Girard was diagnosed with breast cancer in 2006, she decided to take a genetic test to see whether she was also at risk of ovarian cancer.[32] She tested positive. Because the positive test meant that she would have to have her ovaries removed, she asked for a second opinion. She was told this was not possible because Myriad Genetics (refer to Chapter 1, "Hunting for disease genes") owned the patents on both genes and on the tests for measuring the risks of breast and ovarian cancer. This does not mean that the tests devised by Myriad Genetics are not widely available. They are. But the patent prevents others from attempting to develop new tests for these genes, although in point of fact, new tests would be unlikely to provide a second opinion different from the first opinion. However, Myriad Genetics reportedly charges up to $3,000 per test and competitive new tests might bring the price down.[33]

Table 5–2 Cumulative risk of breast cancer by age in women from families with predisposing mutations in either BRCA1 or BRCA2 genes

	Cumulative risk (%)	
Age (years)	***BRCA1***	***BRCA2***
30	3.2	4.6
40	19.1	12.0
50	50.8	46.0
60	54.2	61.0
70	85.0	86.0

Modified after Petrucelli, Nancie et al. "BRCA1 and BRCA2 Hereditary Breast/Ovarian Cancer." *Gene Reviews*. http://www.ncbi.nlm.nih.gov/bookshelf/br.fcgi?book=gene&part=brca1

Genae Girard's inability to get that second opinion formed the basis for a lawsuit filed in federal court in New York on May 12, 2009, by the American Civil Liberties Union and the Public Patent Foundation.[34] The defendants named in the lawsuit were the U.S. Patent and Trademark Office, Myriad Genetics, and the University of Utah Research Foundation. The suit was joined by several national organizations, including the American College of Medical Genetics and Breast Cancer Action, prominent human geneticists, and several women at risk of breast cancer. In total, around 150,000 researchers, pathologists, and laboratory professionals are represented in the suit. Organizations like the American Medical Association and the March of Dimes have also filed friend-of-court briefs in support of the lawsuit. The first hearing was held in federal court on February 2, 2010. On March 29, 2010, U.S. District Court Judge Robert W. Sweet issued a 152-page decision invalidating seven of Myriad's *BRCA 1* and 2 patents on the grounds that the patents had been "improperly granted." The judge argued that the notion that isolating a gene made it patentable was "a 'lawyer's trick' that circumvents the prohibition on the direct patenting of the DNA in our bodies but which, in practice, reaches the same result."

Christopher A. Hansen, Senior National Staff Counsel for the American Civil Liberties Union (ACLU), said that Myriad Genetics was not at fault for asking that the patent be granted, but the patent office was at fault for granting it. When Hansen found that the patent office had authorized that a gene be patented, he couldn't believe it was true. "Patenting human genes is like patenting $e=mc^2$, blood or air," Hansen remarked.[35] A recent paper published by a group at Duke University titled, "Metastasizing Patent Claims on *BRCA1*," argues that the patent is far too broad.[36] For example, it includes over 300,000 short DNA sequences (oligonucleotides) that are found on chromosome 1. This chromosome does not even contain the *BRCA1* gene. It is located on chromosome 17.

Because there are between 3,000 and 5,000 patents on human genes, the outcome of this lawsuit is of great importance to the biotechnology and pharmaceutical industries. If Judge Sweet's decision is upheld on appeal, it will have far-reaching consequences because these patents will become subject to challenge.

Men can develop breast cancer too, but it is rare and usually affects men between the ages of 60 and 70. A much more serious problem for men is prostate cancer. This disease will affect one in six American men over their lifetime.[37] Although there are many prostate cancer susceptibility genes scattered about on different chromosomes, including *BRCA2,* there is no one gene mutation that confers the risks that *BRCA1* and 2 mutations do for breast cancer.[38] Furthermore, this is an ongoing area of research with one group alone recently reporting that they have identified ten new susceptibility loci.[39]

According to the National Cancer Institute, genetic testing for prostate cancer risk is currently not available except as part of a research study.[40] However, deCODE Genetics markets a test for eight susceptibility genes.[41] The company claims that these eight variants combined seem to account for about half the cases of prostate cancer. As we shall see later (see Chapter 10, "The dawn of personalized medicine"), deCODE's prostate cancer test may have saved the company's cofounder from having to grapple with the disease. Nevertheless, the number of prostate cancer susceptibility loci seems to be increasing rapidly, so perhaps deCODE's test will soon contain additional susceptibility loci and capture even more of the risk of a man developing prostate cancer. One must hope so in view of the fact that the utility of the PSA (prostate specific antigen) test for detection prostate cancer has now been seriously questioned.

A malign progression: colorectal cancer

Every year, 160,000 people in the United States are diagnosed with colon cancer and 57,000 of them will die of the disease.[42] Colon cancer is initiated when a benign polyp becomes cancerous (adenomatous). The resulting adenoma then invades the wall of the colon. Up to this point (stages I and II), the cancer can be removed surgically. But in stage III, it metastasizes to the regional lymph nodes. The disease is still curable at this stage with surgery and chemotherapy, but when the cancer reaches distant lymph nodes (stage IV), it is normally incurable.

During the 1980s, Bert Vogelstein at Johns Hopkins School of Medicine began to work out the genetic changes that accompany the physical changes involved in the conversion of a benign polyp to an invasive cancer.[43] Voglestein found that one change involved *TP53*, but he and his colleagues discovered another important gene that was deleted from chromosome 18. This gene was a tumor suppressor that they named deleted in colon cancer (*DCC*). Thus, Vogelstein and colleagues had begun to work out the genetic pathway to colon cancer.

Today, we have a clearer picture of the progressive genetic changes that accompany colon cancer.[44] Genomic instability appears to be a major feature of colon cancer and the disease progresses via either of two pathways. Chromosomal instability (CIN), meaning changes in chromosome number, occurs in around 85% of all colon cancer tumors (see Table 5–3). The *APC* gene plays a role early on in the CIN pathway. This gene is frequently altered in an inherited disease called familial adenomatous polyposis.[45] As you might expect from the name of the disease, people with mutations in the *APC* gene begin to develop multiple benign polyps in their colons as early as their teenage years. Unless the colon is removed, these polyps will become malignant and colon cancer will result.

Table 5–3 Some of the genes involved in the progression of colon cancer

Progression	**Normal Epithelium**	⇒	**Early Adenoma**	⇒	**Late Adenoma**	⇒	**Cancer**	⇒	**Metastatic Cancer**
CIN pathway genes		*APC*		*KRAS*		*TGFBR2* *TP53* *BUB1*			*PRL3*
MSI pathway genes		*APC* *CTNNB1*		*KRAS* *BRAF1*		*TGFBR2* *Others*			

Modified from Grady, William M. "Genomic Instability and Colon Cancer." *Cancer and Metastasis Reviews* 23 (2004): 11–27.

APC is a tumor suppressor gene.[46] The protein encoded by this gene has a variety of functions. The APC protein is involved in regulating how often a cell divides and how a cell attaches to other cells within a tissue, and, most important, it helps to ensure that a cell has the correct number of chromosomes. APC mainly exercises these functions by associating with other proteins involved in cell attachment and signaling. Other genes along the CIN pathway include the *KRAS* oncogene.[47] Like the APC protein, the KRAS protein is involved in regulating cell division. The product of the *BUB1* gene, in contrast, plays an important role in mitosis by ensuring the mitotic spindle is prepared to function properly and that the chromosomes are aligned correctly for cell division.[48] As the progression continues to cancer, *TP53* mutations occur and so do mutations in an oncogene called *TGFBR2*.[49] The latter gene encodes a receptor protein that spans the cell membrane. Signals transmitted through this receptor stimulate various cellular responses, but, most important, one of those signals tells the cell to stop growing and dividing. In short, progression along the CIN pathway involves a series of genetic changes that encourages chromosome instability while promoting uncontrolled cell growth with the endpoint of the pathway being metastasis. When the *PRL3* gene at the end of the genetic progression is overexpressed, it promotes cancer metastasis.

About 15% of colon cancers initiate in a different way, but the progression bears many similarities to the CIN pathway. In this case, the mutations involved affect a set of genes that fix specific mistakes occurring during DNA replication. This repair system is called the mismatch repair system. Examples of the kinds of mismatches this system repairs are when A mistakenly pairs with C rather than T or when G gets it wrong with T rather than its correct mate C. Mutations blocking the function of mismatch repair genes probably lead into the pathway to colon cancer by permitting mutations to occur in *APC*. Except for a few genes, this pathway is quite similar to the CIN pathway.

If the marker for the CIN pathway is chromosomal instability, the one that characterizes the mismatch repair pathway is something called microsatellite instability (MSI). Microsatellites are short, tandemly repeated sequences of two to four nucleotides scattered throughout the genome. For example, $(CA)_n$ is a common repeat sequence. Clusters of these repeats may be amplified from 10 to 100 times. When these microsatellites are extracted from the DNA, they form a characteristic fingerprint that reflects their length and number. When the mismatch repair system is not working, this pattern is altered, often reflecting a change in the number of repeating units.

Smoking, carcinogens, and lung cancer

The explosion of the Unit 4 nuclear reactor at Chernobyl, Ukraine, on April 26, 1986, released massive amounts of radioactive materials into the atmosphere, causing consternation around the world, but especially in Europe, the Baltic countries, and Belarus as the prevailing winds drove the radioactive plume westward. Heavy rains shortly afterward brought much of this material to earth where it entered the ground water. There have been a multitude of studies directed at determining whether this massive radioactive fallout increased the incidence of cancer in nearby countries. An analysis of these results suggests that only childhood thyroid cancer showed unambiguous increases in frequency.[50] Radioactive iodine (^{131}I), generated by the explosion and required for the thyroid hormones thyroxine and triiodothyronine to function, is presumed to be responsible.

Just as Chernobyl shows how radioactive fallout from a nuclear explosion can cause a specific type of cancer, namely thyroid cancer, the carcinogenic compounds present in cigarette smoke cause lung cancer. As Robert Proctor wrote in reviewing the history of tobacco and lung cancer, "tobacco is the world's single most avoidable cause of death."[51]

Tobacco is a New World plant. While Christopher Columbus was in the Bahamas, the local Indians offered him a gift of dried tobacco leaves. They used these leaves in several rituals. Soon thereafter, some of Columbus's men began to smoke the leaves and the habit later became widespread in Europe and elsewhere although usage was sporadic. Cigarettes became popular in the nineteenth century, but the laborious process of rolling them by hand meant that the women who did the rolling were unable to produce more than 200 each day.

Cigarette manufacture received a big boost late in the nineteenth century when automated rolling equipment became available. Clever advertising and promotions exacerbated the habit. Early on, for example, cigarette companies included small cards in each box as a reward.[52] The cards, collectibles that could be traded, depicted famous movie stars, baseball players, and Native American chiefs. The practice was stopped during World War II to save paper, but people still collect the cards and rare cards, like one depicting former Pittsburgh Pirates great Honus Wagner, can garner more than $2 million today. Later, there were posters and magazine advertisements posing attractive men and women with cigarettes. There were also the catchy jingles. "Winston tastes good like a cigarette should," or "Light up a Lucky." If this were not enough, cigarette usage received a big boost in the two world wars when they were dispensed as part of the soldiers' rations. Consequently, after each conflict, an army of fighting men went home as an army of cigarette addicts.

John Hill, an Englishman, first mooted the possible connection between tobacco and cancer in 1761.[53] He observed that excessive use of snuff sometimes caused cancer of the nose. In 1795, Samuel von Soemerring in Germany reported that pipe smokers were prone to cancer of the lip. The connection between pipe smoking and cancer of the lip and mouth was bolstered further in the nineteenth century, but lung cancer was extremely rare before the twentieth century. During the second decade of the twentieth century, physicians started to notice increasing frequencies of lung cancer. In 1912, Isaac Adler suggested a possible connection between lung cancer and smoking, but this connection seemed tenuous at first because the time lag between first exposure and the development of lung cancer is on the order of 20 years. It was only after World War II that lung cancer and smoking were definitively linked. Establishing this link took a long time with the tobacco companies fighting it tooth and nail all the way. The legacy of this battle for the year 2009 was 219,440 new cases of lung cancer in the United States and 159,390 deaths from this disease.[54]

There are at least 55 carcinogens in tobacco smoke of which 20 have been shown to cause lung cancer in at least one animal species.[55] These compounds initiate lung cancer by causing mutations in tumor suppressors like *TP53* and *RB* and also in oncogenes like *KRAS*. Furthermore, there are over 50 cancer-causing compounds in secondhand smoke.[56] According to the American Cancer Society, there were 46 million American smokers in 2001 (the most recent year for which numbers were available).[57] Unfortunately, almost 27% of these smokers were aged 18–24, so the habit seems not to be declining among the young. Cigarette smoking accounts for at least 30% of all cancer deaths. These cigarette-caused deaths are not just from lung cancer, but also from cancer of the larynx, oral cavity, throat, and esophagus. But cancers only account for around half of all smoking-related deaths. Heart disease, emphysema, and stroke are among the life-threatening diseases exacerbated by smoking. Overall, male smokers lose an average of 13.2 years of life and female smokers 14.5 years of life because of their habit. Perhaps the most important lesson lung cancer teaches is that there is usually a very long lag between the time a person begins to smoke and the appearance of the disease. Smoking may seem harmless enough at first, but its consequences can be lethal later on.

Fixing bad blood: lymphoma and leukemia

Lymphoma and leukemia are both cancers of the blood. Lymphomas originate in the lymph system of the body. Lymphocytes are white blood cells that, upon conversion to lymphoma cells, begin to accumulate in masses around the lymph nodes and in other parts of the lymph system. There are three types of lymphocytes. Natural killer (NK) cells are large, granular lymphocytes, whereas B and T cells are small lymphocytes.

Thomas Hodgkin, an English physician, provided the first description of a lymphoma.[58] In 1826, Hodgkin, with a medical degree from Edinburgh in hand, was appointed the first lecturer in morbid anatomy and museum curator at the new Guy's Hospital in London. He performed several hundred autopsies and logged in over 3,000 specimens in the Hospital's Green Book. In 1832, Hodgkin published a paper titled "On Some Morbid Appearances of the Absorbent Glands and Spleen." In his paper, Hodgkin described his postmortem discovery of enlarged lymph nodes or the spleen in the bodies of seven former patients. In 1865, another well-known physician at Guy's Hospital named Samuel Wilks published a paper on the same disease and christened it Hodgkin's disease. Later on, it was discovered that Hodgkin lymphoma is characterized by the presence of multinucleated cells called Reed-Sternberg cells.

Today, lymphomas are classified as either Hodgkin or non-Hodgkin. In 2009, around 601,180 people in the United States were either living with lymphoma or were in remission.[59] They included 148,460 people with Hodgkin lymphoma and 452,720 people having non-Hodgkin lymphoma.

There are many kinds of non-Hodgkin lymphomas with about 85% affecting B cells and the rest T cells or NK cells. Some of these lymphomas are classified as slow growing and others as fast growing. The discussion that follows will focus on the most common fast-growing lymphoma. It is called a diffuse large B-cell lymphoma (DLBCL). The most frequently detected genetic alteration associated with DLBCL is a rather curious one involving a gene called *BCL6*.[60] This gene is not mutated, but instead a piece of a different chromosome becomes attached upstream of the gene (translocation). The translocation attaches regulatory sequences to the *BCL6* gene that perturb its expression.

BCL6 encodes a transcription factor that has many functions. But at least two of these seem to be closely related to its role in lymphoma formation.[61] First, the BCL6 transcription factor is crucial for rapid B cell proliferation. This occurs in sites located within the lymph nodes called germinal centers. Second, this transcription factor appears to inhibit programmed cell death (apoptosis). If these two functions are perturbed, as they are in DLBCL, B cells should proliferate rapidly, but they will not be destroyed by apoptosis. Instead, they will accumulate in masses, a distinctive feature of the disease.

The Leukemia & Lymphoma Society Web site (www.lls.org/hm_lls) has lots of useful information about the respective diseases, including a summary of the various drug therapies currently in use. The one used to treat DLBCL is called R-CHOP and often involves six cycles of treatment. What is R-CHOP? R refers to rituxan, a monoclonal antibody, which binds to CD20, a surface protein on both normal and malignant B cells.[62] Rituxan binding marks these cells for destruction. Stem cells in the bone marrow lack CD20, allowing new, healthy B cells to be regenerated. Over time, these new and normal B cells accumulate while the sheets of cancerous B cells, having been marked for elimination by rituxan, are destroyed.

CHOP refers to four different compounds that interfere with the growth of both normal and malignant cells.[63] Cyclophosphamide (C) and doxorubicin (H) interfere with DNA replication, but in different ways. Vincristine (O) blocks the formation of the microtubule structures required for the assembly of the mitotic spindle essential for cell division, whereas prednisone (P) is an immune suppressant used to slow down the rapid growth of cancerous white blood cells. CHOP is the most common chemotherapy for lymphoma, although there are many other regimens as well. It has been used very effectively for decades to combat highly aggressive lymphomas like DLBCL. To promote white cell regeneration, nulasta is given after each chemotherapy treatment. Nulasta is a genetically engineered human protein called G-CSF that promotes white cell growth.[64] In addition, radiation therapy is used to target those places where cancerous masses have been detected.

In 1960, two cytologists working at different institutions in Philadelphia noted that patients with chronic myelogenous leukemia (CML) had an unusually small copy of chromosome 22.[65] The fact that both investigators hailed from Philadelphia led to the naming of this peculiar chromosome in dubious honor of that city. Later on, it was discovered that the Philadelphia chromosome was one of the two products of a reciprocal translocation that had taken place between chromosomes 9 and 22. This resulted in the fusion of two genes called *ABL1* and *BCR*. *ABL1* is a proto-oncogene. It encodes a tyrosine kinase. Kinases transfer phosphate groups from high-energy donors to various substrates. These kinase-catalyzed phosphate transfers are important cellular-signaling mechanisms. For example, the ABL1 kinase has been implicated in processes of cell differentiation, cell division, cell adhesion, and stress response. When the fusion gene was discovered, the function of the second gene was unknown. Hence, the *BCR* gene was simply named (*B*)reakpoint (*C*)luster (*R*)egion to indicate that the gene fusions could occur at different points within this gene. It now appears that this gene, of which there are several versions depending on the position of the breakpoint, also encodes a kinase. The resulting gene fusion produces an activated kinase that promotes unregulated cell division.

It turned out that the tyrosine kinase domain derived from the *ABL1* gene proved to be the key element in converting the fused gene into an oncogene. If a drug could be developed that blocked the kinase function, it might also do away with uncontrolled proliferation of cancerous white blood cells. In 2001, the FDA approved Gleevec for just this purpose. Although Gleevec also targets 4 of the 90 or so other human tyrosine kinases, this does not seem to cause serious problems. Gleevec was the first anticancer drug to be synthesized with a specific target in mind. Its success has spurred the development of other drugs designed specifically to inhibit the function of other oncoproteins.

As its name implies, CML is a chronic disease.[66] Gleevec treatment cannot cure it, but with treatment most patients having the disease remain symptom free for long periods of time. For treatment of patients who cannot tolerate Gleevec or whose disease is resistant to the drug, the FDA has approved two new drugs, Sprycel and Tarceva.

The big picture

In a book of this kind, it is only possible to touch briefly on a few of the many ways in which gene alterations lead to the initiation and spread of cancer. In 2004, Bert Vogelstein and his colleague Kenneth W. Kinzler published a paper titled "Cancer Genes and the Pathways They Control."[67] It is a short, but highly informative overview of the subject.

According to Vogelstein and Kinzler, there are many fewer pathways to cancer formation than there are genes involved in the process. Because virtually all cancer-causing DNA tumor viruses encode proteins that inactivate both the Rb and p53 proteins, Vogelstein and Kinzler speculate that it may be impossible for a tumor of epithelial origin (carcinoma) to form unless the tumor suppressor pathways controlled by the genes encoding these proteins have been inactivated.

Vogelstein and Kinzler distinguish between solid and liquid tumors, defining the latter group as including lymphomas and leukemias. They note that at least three mutations seem to be required to develop a malignant solid tumor in adults. In contrast, "perhaps liquid tumors don't require as many pathways to be inactivated because their precursor cells are already mobile and invasive, key characteristics that solid tumor cells must develop to become malignant." Interestingly, the events that seem commonly to initiate liquid tumors are chromosome translocations whose fusion leads to conversion of a perfectly normal gene into a cancerous one.

One topic Vogelstein and Kinzler discuss that is not even touched on in this chapter is tumor angiogenesis. This is the method by which tumors attract a blood supply, a critical step in tumor development.

Vogelstein and Kinzler point out that between 1994 and the publication of their article in 2004, "significant gains have been made in cancer therapeutics." They identified three major challenges in the decade ahead: discovery of new genes that play a role in cancer, delineation of the pathway that these genes act upon, and "development of new ways to exploit this knowledge for the benefit of patients." In the end, cancer takes a long time to develop. "It takes 30–40 years for a typical epithelial cell to accumulate the multiple genetic alterations required to progress to metastatic disease." This means that many tumors ought to be detectable when they are still curable by conventional surgery. "Though less dramatic than cures, prevention and early detection are perhaps the most promising and feasible means to reduce cancer deaths within the next decade."

Given that the human body is made up of between 10 and 100 trillion cells (estimates vary) most of which divide, some very rapidly, the surprising thing is not that we get cancer, but rather that it is not even more frequent given that, in a dividing cell population so enormous, mistakes are bound to happen. What is truly remarkable is how efficient cellular repair systems are and how successful cell division control mechanisms are in stopping progression of the cell cycle to allow time for these repair systems to weed out unwanted and noxious mutations. We needlessly put pressure on this remarkable protective system when we smoke, create polluted cities, and spend too much time in the sun.

6

Genes and behavior

In 1978, a woman stopped by the office of Han Brunner, who was studying for his MD at the University of Groningen in the Netherlands, to discuss a problem that was concerning her.[1] It seemed that many of the men in her family were prone to violent outbursts. Brunner wondered why, but before he investigated further, he had to obtain his medical degree. Following that, Brunner obtained a PhD at Nijmegen University, where he found and characterized the gene for myotonic dystrophy for his dissertation research. But he also continued his research on the violent relatives of the woman he had met in 1978.

Finally, in 1993, Brunner and his colleagues reported their findings on the Dutch family.[2] Fortunately for them, it so happened that the woman's maternal great uncle, who was not violent, had kept a record that revealed that nine male relatives exhibited aberrant behavior. Since then, five more men in the family had displayed the disorder. Those studied by Brunner and his colleagues included a man who raped his sister and later, having been confined to a mental institution, chased the warden with a pitchfork. Another brandished a knife in front of his sisters and forced them to disrobe, while a third had attempted to run over his boss after being told that his work was not satisfactory. Two others were arsonists. They all exhibited slight mental retardation, with IQ scores of around 85. The women in the family were neither violent nor mentally deficient.

These findings suggested that a gene on the X chromosome was probably involved. Brunner and his colleagues showed that this indeed was the case and that the gene in question encoded an enzyme called monoamine oxidase A (MAOA). Affected males had a point mutation in the gene that caused a total loss of activity of this important enzyme that plays a central role in the inactivation of four monoamine neurotransmitters found in the brain (serotonin, norepinephrine, melatonin, and dopamine). Although these genetic findings, coupled with their neurological implications, strongly suggested that the absence of MAOA was linked to aggressive behavior, the authors stressed that aggressive behavior varied markedly in severity over time within the pedigree. Naturally, the Dutch findings provoked much interest. Natalie Angier, writing in the *New York Times* under the headline "Gene Tie to Male Violence Is Studied," made a correct prediction when she prophesied that the discovery of this genetic defect was "likely to rekindle the harsh debate over the causes of criminal and abnormal behavior."[3]

The gene defect unique to the Dutch pedigree proved extremely rare, but later on, a polymorphism was found upstream of the *MAOA* gene that has a profound effect on its expression.[4] This was a repeated sequence of 30 base pairs. Although there were several variants, three and four repeats were by far the most common. Expression of the *MAOA* gene was considerably lower for the "short" (three repeats) version of the gene than the "long" (four repeats) version.

This discovery has been the basis for a whole series of investigations on antisocial behavior and MAOA levels. A key contribution was made by Avshalom Caspi and Terri Moffit, clinical psychologists at the University of Wisconsin and King's College London, and others in 2002.[5] These investigators decided to examine whether *MAOA* gene expression could be correlated with aggression among children who had experienced maltreatment. The children in question had grown up by the time Caspi and his colleagues conducted their study, but they had been assessed repeatedly over the intervening years since they were part of a cohort of 1,037 babies who had been born in Dunedin, New Zealand, between April 1, 1972, and March 31, 1973. During the "assessment phase," all members of the Dunedin Study are brought back to the University of Otago where the study is located and their physical and mental health is examined.[6] Long interviews are part of the process.

Caspi and colleagues focused on male children because they have a single X chromosome and, hence, but one copy and one form of the

MAOA gene. They determined whether each individual had the long or short form of the gene, using this as a proxy for MAOA activity. They found that the interaction between maltreatment and antisocial behavior, using four recognized measures of the latter, was significantly weaker among males with high MAOA activity than it was in males with low MAOA activity. Furthermore, men with low MAOA activity who had been maltreated were more likely to be convicted of a violent crime than men with the same low activity who had been brought up normally. In contrast, there was no significant increase in risk of violent crime among maltreated men with high MAOA activity. One important difference between the Caspi study and Brunner's results was that the men profiled in the Caspi report had normal IQs, so lower IQs could not be invoked as a reason for frustration, leading ultimately to violence. Additional support for Caspi's conclusion was soon forthcoming when a group in Virginia confirmed their results using a sample of boys from the Virginia Twin Study for Adolescent Behavioral Development and somewhat different criteria for measuring conduct disorder.[7]

In 2004, journalist Ann Gibbons wrote up the highlights of the annual meeting of the American Association of Physical Anthropologists held in Tampa, Florida, for *Science* magazine.[8] Gibbons reported that a team of geneticists led by Tim Newman at the National Institute of Alcohol Abuse and Alcoholism had found the short variant of the *MAOA* gene in chimpanzees, gorillas, and many other primates. However, Newman also discovered that the short variant was found only in Old World (Asian and African) monkeys, but not in New World (South American) species. Hence, the short variant appeared to have arisen after Old and New World monkeys separated some 25 million years ago. Newman suggested that there must be a reason why the short variant was present at such a high frequency in human and Old World primate populations. "Bold, aggressive males might have been quicker to catch prey or detect threats."

Gibbons titled her article "Tracking the Evolutionary History of a 'Warrior' Gene." And, of course, the name caught on for the short version of the *MAOA* gene.

But later, Newman and colleagues proceeded to perform the rhesus monkey equivalent of the Caspi study, but with somewhat surprising results.[9] Monkeys reared without mothers were supposed to be the equivalent of badly treated human children, but the monkeys with the least amount of MAOA activity also showed the lowest levels of aggression. When their mothers reared these monkeys, they exhibited the highest levels of aggression of any group. Conversely, the monkeys with high levels of MAOA activity showed the lowest levels of aggression when reared by their mothers, but aggression increased markedly when they were reared without parents. As Newman and colleagues point out, the situations in the Caspi and their studies are somewhat different. First, distinct species were involved. Second, mothers versus no mothers were compared in the Newman experiments, whereas in Caspi, all the males in question had parents, some of whom were bad. Nevertheless, the differences in results between the two studies are important because they illustrate why one must be careful in extrapolating the results of animal studies, even when primates are involved, to predict the possible outcomes of human studies conducted in a similar fashion.

Warrior gene studies turned sinister when a group led by Kevin M. Beaver of Florida State University's College of Criminology and Criminal Justice linked the low activity form to gang membership.[10] This caught the attention of the national media. Reporter Kathleen Kingsbury titled an article in *TIME* magazine "Which Kids Join Gangs? A Genetic Explanation."[11] But of even greater consequence was the use of the warrior gene argument to convince a jury that a convicted murderer did not deserve the death penalty.

On the evening of October 16, 2006, Bradley Waldroup was drinking in his trailer home in the mountains of Tennessee while awaiting the arrival of his estranged wife and their four children for the weekend.[12] An obviously concerned Penny Waldroup arrived with her children and her friend Leslie Susan Bradshaw having left word with another friend to call the police if they did not return by a specified time.

When Penny Waldroup told her husband that she and her friend were leaving, an argument ensued. In a rage, Brad Waldroup shot Bradshaw eight times and then cut her head open with a sharp object. Next, he chased his wife with a machete, chopping off a finger and cutting her again and again. When the police arrived, Bradshaw was dead, Penny Waldroup was badly injured, and there was blood all over the trailer, including the walls, the carpet, and even on the bible that Brad Waldroup had been reading while awaiting the arrival of his wife and children.

Prosecutors charged Waldroup with the felony murder of Bradshaw, a death penalty offense, and with the attempted first-degree murder of his wife. Prosecutor Cynthia Lecroy-Schemel said that there "were numerous things he did around the crime scene that were conscious choices. One of them was [that] he told his children to 'come and tell your mama goodbye,' because he was going to kill her. And he had the gun, and he had the machete." The case was pretty cut and dried as even Waldroup admitted. He claimed to have "snapped" and said, "I'm not proud of none of it."

Because the defense could not dismantle the evidence, they had to take a different tack. As attorney Wylie Richardson put it, "It wasn't a *who done it*? It was a *why done it*?" To try to get an answer on "why," Richardson asked forensic psychiatrist William Bernet at Vanderbilt University to give Waldroup a psychiatric examination. Bernet did so, but he also took a blood sample as one of his interests was the relationship between MAOA and criminal behavior and he was quite familiar with the work of Caspi and colleagues.[13] In fact, by the time of the Waldroup trial, Bernet and his colleagues had tested some 30 criminal defendants, the majority having been charged with murder.[14]

The prosecution objected strongly to Bernet's being allowed to testify, but the judge allowed his testimony. Bernet and his colleagues found that Waldroup indeed had the warrior gene. Because he had also been abused as a child, Bernet said that Waldroup's "genetic makeup, combined with his history of child abuse, together created a vulnerability that he would be a violent adult." Bernet was careful to point out that he hadn't said, "these things made him become violent, but they certainly constituted a risk factor of vulnerability."

Prosecutor Drew Robinson believed the genetic evidence was "smoke and mirrors" designed to confuse the jury. He called in a psychiatrist named Terry Holmes, the Clinical Director of Moccasin Bend Mental Health Institute in Chattanooga, Tennessee. Holmes belittled Bernet's interpretation saying it was too early to try and make such a connection in a court of law. Holmes believed that because Waldroup was "intoxicated and mad" he "was gonna hurt somebody. It had little to nothing to do with his genetic makeup."

After deliberating on the Waldroup case for 11 hours, the jury convicted Waldroup of voluntary manslaughter and attempted second-degree murder. Drew Robinson, the prosecutor, couldn't believe his ears. "I was just flabbergasted. I did not know how to react to [the verdict]." In explanation, juror Sheri Lard said the warrior gene was just one piece of evidence. Some attached importance to it, but for others it was less significant. Where it was decisive for the jury was in deciding whether to find Waldroup guilty of murder and subject to the death penalty. For juror Debbie Beaty, the genetic evidence was persuasive. "A diagnosis is a diagnosis, it's there," she said. "A bad gene is a bad gene."

Beaty's statement explains why the prosecution was so anxious to prevent Bernet's testimony. Bernet was an expert where the warrior gene was concerned, but Holmes clearly was not. This fact was not lost on the jury who, given the nature of the evidence, concluded that Waldroup's actions might have been beyond his control.

On a lighter note, perhaps, political scientists and economists got into the warrior gene act using game playing as a measure of aggression.[15] In this case, they measured the willingness of male subjects to pay to administer hot sauce to someone who had taken money from them. It turned out there was little difference between low and high MAOA activity groups when only 20% of their money was taken, but the warrior gene participants administered hot sauce more aggressively than their calmer colleagues after losing 80% of their money.

And now, we have the *MAOA* gene predicting credit card debt according to Jan Emmanuel Neve, a graduate student at the London School of Economics, and James H. Fowler, a political scientist at the University of California, San Diego.[16] They report that having one copy of the low-efficiency variant raises the average likelihood of having credit card debt 7.8%, whereas having two copies increases this number to 15.8%. Their numbers also suggest there may be a sex bias because men have only one copy of the *MAOA* gene, whereas women have two. This is the kind of observation that titillates the media and was picked up by *Newsweek* and the *Wall Street Journal's Smart Money*.[17]

Race also briefly raised its ugly head with regard to the warrior gene. In August 2006, Dr. Rod Lea, a genetic epidemiologist at the

ESR Kanepuru Science Centre in Porirua, New Zealand, caused a furor, particularly Down Under, when he presented a report titled "Tracking the Evolutionary History of the Warrior Gene in the South Pacific" at the International Congress of Human Genetics held in Brisbane, Australia. As the *Sydney Morning Herald* reported on August 9, Lea told Congress participants that the Maori had a "striking over-representation" of the warrior gene and that this went "a long way to explaining some of the problems Maoris have."[18] "Obviously, this means they are going to be more aggressive and violent and more likely to get involved in risk-taking behaviour like gambling." Lea added the caveat that he thought nongenetic factors might also play a role, but then he went on to link the gene to smoking and drinking among the Maori.

Lea's paper provoked an eruption that he hastened to quell in an editorial in the *New Zealand Medical Journal* where he attempted to put a positive spin on his findings. Lea argued that his main point was that the warrior gene had been selected in the Maori in connection with their breathtakingly risky canoe migrations over vast stretches of ocean not knowing whether they would find land or not. Lea, and his colleague Geoffrey Chambers, pleaded that they were ethical researchers who had received far too much negative media hype.

The most thoughtful contribution to the whole debate over the warrior gene and Maori aggression came from a distinguished Maori scientist named Gary Raumati Hook.[19] He aimed to assess the "truth" of the warrior gene hypothesis and what it might mean for the Maori. Hook summarized the many negative behaviors that have been linked to the warrior gene (see Table 6–1). But as he pointed out, "possession of these traits would pose severe disadvantages especially in the ancient world of the voyaging and war-like Polynesians." Furthermore, "in the case of Maori domestic violence blaming nature as opposed to nurture simplifies the problem because it implies that Maori has no one to blame but himself for his own condition. Contributions to racial stereotyping by trained scientists are unethical and scandalous."

In fact, the problem with the whole warrior gene hypothesis is that the low-activity variant is common in most populations (see Table 6–2), yet the majority of people with this variant do not exhibit bad behavior. Because it is so abundant, it probably does have some selective advantage. The variety of anomalies that have been associated with the warrior gene variant (refer to Table 6–1) show just how difficult it is to sort out behavioral and more complex conditions involving many genes. On top of that, there are environmental conditions that play an important role. Studies like those of Caspi et al.[20] illustrate the importance of combining a specific environmental trigger (child abuse) with the warrior gene in causing bad behavior.

Table 6–1 Behavioral disorders reported to be associated with low MAOA activity*

Disorder	Reference**
Personality disorders	19
Antisocial behavior	5,7
Violence and risk taking	19
Impulsive aggression	19
Mental disorders	19
Credit card debt	16,17
Obesity	19
Propensity to join gangs	10
Behavioral aggression in game playing	15

*Modified from Hook, G. Raumati. "'Warrior genes' and the disease of being Maori." MAI Review 2 (2009). Target article.
**See chapter references at end of book.

Table 6–2 Estimated frequencies of the short (3-repeat) version of the MAOA gene in different ethnic groups

Ethnic group	Frequency of 3-repeat (%)	Number of chromosomes sampled
Caucasian (males)	34	2,382
Chinese (males)	77	55
African (both sexes)	59	52
Hispanic (both sexes)	29	27
Pacific Islander (both sexes)	61	50
Maori males (at least one Maori parent)	56	46

Adapted from Lea, Rod, and Geoffrey Chambers. "Monoamine Oxidase, Addiction, and the 'Warrior' Gene Hypothesis." *The New Zealand Medical Journal* 120 (March 2, 2007). http://www.nzma.org.nz/journal/120-1250/2441.

Fragile X syndrome

Around one in 4,000 men and one in 8,000 women suffer from fragile X syndrome, making it the most common form of inherited mental retardation.[21] The unraveling of this hereditary disease began with a 1943 paper by British neurologist James Purdon Martin and geneticist Julia Bell describing a sex-linked mental defect.[22]

Bell was well into a long and remarkable career by the time she published her paper with Martin.[23] In the early 1900s, she spent two years studying mathematics at Cambridge, but as a woman she was unable at the time to obtain a degree from that institution so she transferred to Trinity College, Dublin, as many women did, to complete her studies. Meanwhile, Karl Pearson, the great statistician, and a disciple of Francis Galton, was establishing the Galton Laboratory in National Eugenics, at University College London. He would also be the first occupant of the Galton Professorship at that institution. In 1908, Bell began working for Pearson as a statistician and in 1909, he initiated the *Treasury of Human Inheritance.* This compendium on human genetic disorders, published between 1909 and 1958, was one of the first such undertakings in the field and Bell was responsible for writing most of its sections.

Martin and Bell had established that the fragile X condition, also known as Martin-Bell syndrome, was sex-linked, but it was Herbert Lubs who observed, in another pedigree of sex-linked mental retardation, that there was a constriction near the end of the long arm of the X chromosome in cultured lymphocytes.[24] Lubs's results proved irreproducible initially because the constriction, or fragile site, was not evident using cell culture media different from those used by him. But once this had been realized, the constriction that Lubs had described became evident once more.

The gene responsible for fragile X syndrome was isolated in 1991 and named *FMR-1* for fragile X mental retardation-1.[25] The disease proved to be caused by overamplification of a CGG repeat sequence. That is, it was a trinucleotide repeat disease like Huntington's chorea (see Chapter 1, "Hunting for disease genes," and Table 1–1). Normal individuals have 5–54 repeats of this arginine-specifying codon. People with 55–200 repeats are referred to as having a premutation, whereas those with more than 200 repeats have the full mutation with the gene becoming completely inactive due to epigenetic methylation of cytosine residues in the repeat sequence.[26] The inactivation of *FMR-1* means that its protein FMRP is not made. Impairment of brain function then occurs, probably in part because synaptic connections between at least some neurons do not form properly.

Fragile X syndrome is dominant, meaning that the disease is expressed in women as well as in men. People with the premutation are usually unaffected.[27] However, in women with the premutation, the repeat sequence in the affected X chromosome may amplify during oogenesis so that some of their children may express the full-blown disease. Babies with fragile X syndrome seem normal during the first year, but they experience delays in speaking and language development with lapses in short-term memory and problem-solving ability appearing later on. Autism is often associated with fragile X syndrome, although individuals with fragile X syndrome only account for 2%–6% of autism cases.[28] IQ tests reveal mild to severe mental retardation in men (IQs of 20 to 70), but women and less-affected men may have IQs approaching 80. Characteristic physical changes also occur in people with fragile X disease, so they tend, for example, to have long, thin faces and prominent ears. Life expectancy is normal.

The gay gene controversy

In 1991, J. Michael Bailey and Richard C. Pillard reported the results of a twin study designed to examine the heritability of homosexuality.[29] They found 52% of identical twins in their sample were both homosexual, but this dropped to 22% among nonidentical twins, and 11% among adoptive brothers. Because identical (monozygotic) twins possess identical genomes, whereas nonidentical (dizygotic) twins are no more alike genetically than brothers and sisters, the results Bailey and Pillard reported strongly suggested that homosexuality was at least partially genetic in nature.

Meanwhile, Simon LeVay published an anatomical study in *Science* magazine that same year in which he reported that the anterior hypothalamus in men was twice as large as the same organ in women and homosexuals.[30] This region is believed to control sexual activity. He had made this comparison because an earlier study had demonstrated that this portion of the hypothalamus was larger in males than in females.

Based on these studies, *Newsweek* ran a 1992 article on homosexuality titled "Born or Bred?"[31] The article was a well-balanced review of these studies, the reasoning of their critics, and opinions of experts in the area and of several gays. At the end of their article, the author turned to Evelyn Hooker for an answer. In the 1950s, Hooker showed that it was impossible to distinguish between heterosexuals and homosexuals on the basis of psychological testing. Hooker asked, "Why do we want to know the cause? It's a mistake to hope that we will be able to modify or change homosexuality...If we understand its nature and accept it as a given, then we come much closer to the kind of attitudes which will make it possible for homosexuals to lead a decent life in society." Hooker was 84 when she was interviewed, but as the authors of the *Newsweek* article said, "The rest of us could do worse than listen to her now."

Despite Hooker's sensible advice, the question of whether being gay was genetically determined was in the news in a big way the following year. On July 19, 1993, President Bill Clinton announced a new policy regarding gays in the military.[32] Recruits would not be asked about their sexual orientation and, if they were homosexuals, would be obliged not to reveal this fact. Three days earlier, a paper had appeared in the journal *Science* claiming that a region on the long arm of the X chromosome called Xq28 contained a genetic variant that predisposed men to be homosexuals.[33] This report naturally received wide coverage in the press. The senior author was Dean Hamer, an investigator at the National Institutes of Health. Hamer did not claim to have identified a gay gene per se, but only a chromosome region that appeared to contain such a gene.

John Maddox, editor of *Nature,* discussed the reaction of the British press to Hamer's paper in an editorial titled "Wilful public misunderstanding of genetics."[34] Maddox worried about "the tendency of even sober-sided newspapers to overdramatise discoveries only, afterwards, to complain that they have been misled. Even a casual reading of the original article will reveal a commendable list of caveats." In contrast, the American newspapers generally exhibited remarkable restraint. The titles of most American newspaper articles on Hamer's report tended to be variants on that in the *New York Times*: "Report Suggests Homosexuality Is Linked to Genes."[35] Although the *Philadelphia Daily News* picked up an AP report titled "Scientists ID Gene Common to Gays,"[36] most papers pointed out that Hamer only claimed to have found a region on the X chromosome that was linked to homosexuality. This was also true of a July 26 article "Born Gay" in *TIME* magazine. That article got to the crux of the problem by pointing out that "if homosexuals are deemed to have a foreordained nature, many of the arguments now used to block equal rights would lose force."[37]

Several articles quoted Elliot Gershon, Chief of the Clinical Neurogenetics branch at the National Institute of Mental Health, who called Hamer's report "a major breakthrough in behavioral genetics," but cautioned that the study must be replicated for credibility.[38] But Richard Knox, writing in the *Boston Globe,* realized that, if Hamer was right, there could be serious consequences.[39] He pointed out that "a prenatal test could conceivably tell parents if a fetus carried a gene that might predispose the child to homosexuality so they could choose abortion."

Nevertheless, there were skeptical scientists. Evan Balaban, then at Harvard University, and Brown University professor Anne Fausto-Sterling raised several questions, including "the lack of an adequate control group."[40] Fausto-Sterling and Balaban argued that Hamer should have looked for Xq28 DNA markers in the nonhomosexual brothers of the gay men analyzed in the study because "chromosomes without the markers should show up in non-homosexual brothers." Hamer and colleagues responded that they did have a proper control group of "142 randomly selected brothers" and that "because homosexuality is a stigmatized trait, individuals who identify themselves as gay are expected to give more reliable information about any homosexual behavior than those who identify themselves as straight."[41] But as Fausto-Sterling and Balaban had pointed out, if the mother had been a carrier of the Xq28 markers, then the straight brothers should have lacked them. Then statistical geneticist Neil Risch, who had originated one of the statistical methods used by Hamer, raised technical questions about Hamer's analysis.[42]

In 1994, Hamer publicized his findings in a popular book called *The Science of Desire* in which he detailed his hunt for the gay gene.[43] To add credibility to his case, in 1995, Hamer's group reported a follow-up study with a different group of subjects that again indicated linkage of homosexuality to Xq28 in men, but not in women.[44]

But the doubts persisted. On June 25, 1995, John Crewdson of the *Chicago Tribune* broke a story titled "Study on 'Gay Gene' Challenged Author Defends Findings Against Allegations."[45] Apparently, a postdoctoral fellow who was an author on the study raised questions about the work and was ordered to leave Hamer's laboratory shortly thereafter. Crewdson wrote that the Office of Research Integrity was looking into the accusations. In the same article, Crewdson reported that Dr. George Ebers, a neurogeneticist from the University of Western Ontario, had searched unsuccessfully "for a Hamer-style genetic link to homosexuality in more than 50 pairs of gay Canadian brothers."

On February 7, 1999, the *Boston Globe* ran an article by Matthew Brelis titled "The Fading Gay Gene."[46] Brelis cited the still unpublished Canadian study, but he also quoted Dr. Richard Pillard, who was involved in the study of twins and sexuality, as pointing out that while a gay man's fraternal twin has a 22% chance of being gay, the frequency rises to only 50% among identical twins. Concordance should be 100% if genes were everything. As Brelis wrote, "interest in—and enthusiasm for—the 'gay gene' research has waned among activists and scientists alike." On April 23, 1999, the Canadian study appeared in *Science.*[47] It showed no linkage of homosexuality to DNA markers at Xq28; Neil Risch was one of the authors. Hamer, of course, disagreed, citing possible problems with the way in which the Ontario group selected its subjects. The *New York Times* reported the new finding the same day as did the *Washington Post.* "I think the jury is still out," the *Post* quoted Elliot Gershon, now Chairman of Psychiatry at the University of Chicago, as saying. "All of theses studies may be essentially accurate, and until we have hundreds of gay men enrolled in studies we are not going to get a firm answer." Meanwhile, Hamer seems quietly to have put his X-linked gay gene aside. In 2005, he and his colleagues reported on a genome-wide scan for genes related to male homosexuality.[48] This time, they found suggestive evidence for genes that might be involved in chromosomes 7, 8, and 10, but no linkage to Xq28 in the full sample.

Bipolar disease and the biological clock

The prevalence of bipolar disorder or manic depression was previously thought to be around 1% in the population, but more recent diagnoses indicate that it could be as high as 5%.[49] Family, twin, and adoption studies all suggest that bipolar disorder has an important genetic component. Furthermore, a host of studies has pointed fingers at a great many different genes. Neil Risch and David Botstein addressed this point in a 1996 paper titled "A Manic Depressive History." As they wrote, "the recent history of genetic linkage studies for this disease is rivaled only by the course of the disease itself. The euphoria of linkage findings being replaced by the dysphoria of nonreplication has become a regular pattern, creating a roller coaster-type existence for many psychiatric genetics practitioners as well as their interested observers."[50] Risch and Botstein went on to make specific suggestions for tightening up the criteria for establishing genetic linkage. Yet a 2006 review of linkage and genome-wide association studies implicated sites on 19 chromosomes as being associated with bipolar disease with the best evidence for linkage for one and sometimes two sites on 12 chromosomes, still a large number.[51] Numerous candidate genes were also identified.

What are we to make of all this? Some of the genes are of the kind one might expect. Their protein products are involved in breaking down neurotransmitters that conduct signals from one neuron to another (*COMT*), encode a receptor for the neurotransmitter dopamine (*DRD4*), or specify a neuronal structural protein (*DISC1*) with the latter gene also implicated in schizophrenia (see the following section). But the most interesting genes, ones where a coherent story may be coming together, involve the biological clock.

The biological or circadian (from the Latin *circa* for *about* and *dies* for *day*) clock is the roughly 24-hour cycle of biological, physiological, and behavioral processes found in virtually all living things.[52] These rhythms are endogenous, but they are adjusted to environmental changes by external cues, the most important of which is light. Because oscillations in the clock are not precisely 24 hours, it is reset daily.

In mammals, the biological clock is located in the paired suprachiasmatic nuclei (SCN) of the anterior hypothalamus of the brain.[53] Light makes its way to the SCN via the eye and the retinohypothalamic tract. The SCN interprets the information on day and night length obtained from the retina. This information is transferred to the pineal gland, located on the epithalamus, and the pineal gland secretes the hormone melatonin in response. Secretion of melatonin increases at night and declines during the day. Melatonin is involved in the diverse bodily functions (e.g., sleep-wake cycle) connected to the circadian cycle. Melatonin is sometimes used to ameliorate jet lag because it will essentially reset the clock. So if melatonin is taken on an eastbound flight to Europe before endogenous secretion of melatonin begins, it will reset the clock to an earlier time.

The genetic control of the biological clock has been worked out largely in the fruit fly *Drosophila,* but the genetic homologs have been identified in mice and people as well (see Table 6–3). In *Drosophila,* expression of the period (*per*) and timeless (*tim*) genes is stimulated in the morning. Their protein products are transcriptional regulatory proteins. They accumulate in the cytoplasm during the day. In the evening, they complex with each other, enter the nucleus, and turn off their own expression. The protein products of the cycle (*cyc*) and clock (*clk*) genes also play a role. When they complex to form dimers, they stimulate *per* and *tim* expression.

Table 6–3 Homologous genes in Drosophila and mice that play a role in circadian clock regulation

Drosophila	Mouse
Period (per)	mPeriod1
	mPeriod2
	mPeriod3
Timeless (tim)	None
Time-out	mTimeless
Cryptochromes (Cry)	mCry1
	mCry2
Clock (clk)	mClock
Cycle (cyc)	mBmal (MOP3)
	mBmal2 (MOP9)
Double-time	casein kinases 1 epsilon (TAU)

Adapted from Rivkees, Scott A. "Circadian Rhythms—Genetic Regulation and Clinical Disorders." *Growth: Genetics & Hormones* 18 (2002): 1–6.

In mice and people, the expression of the three *per* genes and the genes encoding the blue light photoreceptors cryptochromes 1 and 2 (*cry 1* and 2), which also play a role in the circadian cycle, is driven by transcriptional activating factors encoded by the *clk* and *bmal1* genes. The latter is the homolog of the *cyc* gene of *Drosophila*. The products of the *clk* and *bmal1* genes form an active complex in the same way as their *Drosophila* homologs.

With this background on the circadian cycle and its genetic control, let us turn to its possible relationship to bipolar disease. Colleen McClung, at the University of Texas Southwestern Medical Center, and her colleagues have characterized the behavior of mice with a mutation in the *clk* gene that prevents the CLOCK protein from stimulating transcription of the *per* and *cry* genes.[54] They find that the behavioral profile of the mutant mice partially mimics bipolar disease. For example, during the manic phase of the disease, affected people generally experience an increase in energy and a decreased need for sleep. They can also be more irritable and often abuse substances like alcohol or cocaine. In contrast, the depressive phase of the disease is often characterized by persistent feelings of sadness, anxiety, guilt, anger, isolation, or hopelessness among other things.

Mice with the *clk* mutation are hyperactive, sleep less, and exhibit an increased response to cocaine. Furthermore, lithium treatment alleviates the symptoms just as it does in the case of human bipolar disease. However, the mouse model only replicates the behavior characteristic of mania; it does not replicate the mood oscillations characteristic of bipolar disease. Even though the results obtained by McClung and her colleagues incompletely model bipolar disease, they suggest that focusing on genes affecting the behavior of the circadian clock will probably be a profitable avenue for future research on bipolar disease. This is already happening. One study, for example, reported that SNPs associated with three genes including *clk* were significantly correlated with bipolar disease[55] while another paper documented an association of the *tim* gene with depression and sleep disturbance.

DISC1 and schizophrenia

Schizophrenia is fairly common, with a median lifetime prevalence of 0.7%–0.8% with the malady usually appearing in adolescence or in early adulthood.[56] Like bipolar disease, schizophrenia involves important environmental as well as genetic factors. One of the genes that clearly plays a role in schizophrenia when broken is called *DISC1*.

In July 1990, a group of scientists based in Edinburgh reported on a large Scottish pedigree in which many members suffered from schizophrenia or other mental disorders.[57] These diseases were associated with a translocation between chromosomes 1 and 11. Subsequently, the translocation was found to have disrupted two genes on chromosome 1. They were named *DISC1* and *DISC2* for disrupted in schizophrenia 1 and 2.[58] *DISC1* encodes a protein while *DISC 2* specifies a noncoding RNA molecule (antisense) that could regulate expression of *DISC1* by binding to *DISC1* messenger RNA and preventing its translation.[59] So far, attention has been focused almost entirely on *DISC1*.

Individuals with the translocation in the original Scottish pedigree showed a variety of symptoms ranging from schizophrenia to normality (see Table 6–4). In fact, over one-quarter of individuals having the translocation seemed to be normal. In this regard, it is important to note that individuals with the translocation do not lack the DISC1 protein completely because a good copy of the undisrupted gene is also present on the unaffected copy of chromosome 1. In fact, the DISC1 protein is present in about half the normal level in tissue culture cell lines containing the translocation.[60] So, it appears that other environmental factors as well as differences in genetic backgrounds between individuals probably determine the behavioral differences between the various individuals in the pedigree.

Table 6–4 Psychiatric diagnoses of individuals carrying the translocation disrupted *DISC1* and *DISC2* genes

Diagnosis	Number affected
Schizophrenia	7
Bipolar affective disorder	1
Recurrent major depression	10
Adolescent conduct disorder	2
Anxiety, alcoholism, minor depression	1
Unaffected	8

From Millar, J. Kirsty et al. "DISC1 and DISC2: Discovering and Dissecting Molecular Mechanisms Underlying Psychiatric Illness." *Annals of Medicine* 36 (2004): 367–378.

Although the translocation seen in the Scottish pedigree has not been seen in other families with a history of schizophrenia, other mutations have been found in *DISC1* that affect behavior.[61] For example, a deletion mutation in the *DISC1* gene has been characterized in one pedigree.[62] The father is a carrier, but he is asymptomatic, whereas three of his children, who are also carriers, suffer from schizophrenia or schizoaffective disorder. What is surprising is that the mutation is absent in three other siblings, two of whom suffer from major depression and one who has schizotypal disorder, a malady that mimics some of the symptoms of schizophrenia.

Expression of the DISC1 protein is developmentally regulated and it interacts or complexes with a number of proteins involved in neurodevelopment (see endnote 63 for more details).

Genes and alcoholism

In 1910, Karl Pearson, Director of the Galton Laboratory of Eugenics at University College London, and Miss Ethel Elderton, the Francis Galton Scholar in the laboratory, published a sensational memoir on the heritability of alcoholism.[64] They reported that in data they had analyzed from Manchester and Edinburgh, there was no evidence that parental alcoholism had any influence on the mental capacity or physical characteristics of their children. However, they strongly suspected that alcoholism did have a hereditary basis so that "the problem of those fighting alcoholism is one with the fundamental problem of eugenics."

Despite this qualification, members of the Society for the Study of Inebriety as well as those who belonged to the Eugenics Education Society went ballistic. Montague Crackenthorpe, Chairman of the Eugenics Education Society, spoke for many in his letter to the *Times* when he wrote "to those who are familiar with eugenic... research the Report causes no surprise at all. It simply confirms their belief that, serviceable as biometry is in its proper sphere, it has its limitations, and that a complex problem such as that of the relation of parental alcoholism to offspring is quite beyond its ken."

So people have been thinking about heredity and alcoholism for a long time, but, despite much literature on the subject, with one exception, it is only in the last ten years or so that the complex relationship between genes and alcohol has begun to be unraveled. That exception relates to the fact that East Asians like Chinese and Japanese often become very flushed when they have been drinking alcohol.[65] The alcohol-related flush has been the subject of research for decades and is now well understood. The reason why Asians flush after drinking alcohol is that they have increased levels of a breakdown product of alcohol called acetaldehyde. The *ALDH2* gene encodes the enzyme that degrades acetaldehyde (aldehyde dehydrogenase). The *ALDH2K* variant increases turnover of the enzyme so there is less of it around and, hence, the degradation of acetaldehyde proceeds more slowly, allowing the compound to increase in the bloodstream.[66] This variant is found at frequencies of up to 50% in Asian populations. It is much less common among Caucasians.

Twin studies have indicated repeatedly that the heritability of alcohol dependence is in the range of 50% to 70%.[67] Studies of American Indians in the Southwest and also with Caucasian populations have led to the conclusion that *GABRA2* is a susceptibility gene for alcohol dependence. This gene encodes a protein subunit of the receptor for a neurotransmitter called GABA (gamma-aminobutyric acid). GABA, the main inhibitory neurotransmitter in the brain, helps to induce relaxation and sleep. However, the association with alcohol dependence appears to be modest and not well understood. This is consistent "with a model whereby multiple genes of small effect contribute to vulnerability to alcohol dependence."[68]

Once again, sorting out the roles of nature and nurture for this complex trait is an ongoing challenge. Members of the Collaborative Study on the Genetics of Alcoholism (COGA) consortium have carried out much of the important work linking genetic variants to alcohol dependence.[69] COGA investigators initially carried out genetic linkage studies to identify risk-factor genes for alcoholism. In the process, they assembled a collection of over 300 extended families totaling more than 3,000 individuals in which alcoholism is a serious problem. COGA investigators have profiled these families in detail for clinical, neuropsychological, electrophysiological, biochemical, and genetic data.

The World Health Organization estimates that there are 2 billion alcohol users, 1.3 billion tobacco users, and 185 million illicit drug users worldwide.[70] One of the most interesting developments in recent years is the discovery that genes, variants of which may be involved in alcohol dependence, may also play a role in addiction to other substances, including nicotine and other drugs (see Table 6–5). For instance, *CYP2A6*, a known risk-factor gene for nicotine dependence (refer to Chapter 4, "Susceptibility genes and risk factors"), also appears to be a risk-factor gene for alcohol dependence. These findings are consistent with the interpretation that different substances often trigger addiction through one or more common pathways. If this proves to be the case, it should greatly simplify the process of untangling the genetics of addiction.

Table 6–5 Genes associated with at least one drug addiction

Gene	Biological function	Drug association
5-hydroxytryptamine transporter (*5HTT*)	Neurotransmitter transport	Alcohol, cocaine, heroin, nicotine, methamphetamine
Cytochrome P450, family 2, subfamilyA, polypeptide 6 (*CYP2A6*)	Oxidation, reduction	Alcohol, nicotine
Dopamine transporter (*DAT1*)	Neurotransmitter transport	Alcohol, cocaine, heroin, nicotine, methamphetamine
Dopamine receptor 2 (*DRD2*)	Synaptic transmission dopaminergic	Alcohol, cocaine, heroin, nicotine
Interleukin 10 (*IL10*)	Cytokine activity	Alcohol
Brain-derived neurotropic factor (*BDNF*)	Regulation of synaptic plasticity	Alcohol, nicotine, cocaine, methamphetamine

Adapted from Li, Ming D., and Margit Burmeister. "New Insights into the Genetics of Addiction." *Nature Reviews: Genetics* 10 (2009): 225–231.

7

Genes and IQ: an unfinished story

Why would IQ be included in a book on genes and disease? The reason is that low IQ has often been equated with poor heredity and treated as if it were a disease. From early in the last century until today, those scientists with a hereditarian (as opposed to an environmental) bent have argued that cognitive deprivation is largely the result of a less-than-satisfactory genetic background. This chapter briefly outlines the fraught history relating genes and environment to intelligence. In the first part of the last century, the presumed genetic basis of "feeblemindedness," as it was called, served as the "scientific" basis for the involuntary sterilization statutes passed by many states in America, at the provincial level in Alberta and British Columbia in Canada, and at the federal level in the Scandinavian countries and Nazi Germany. The notion that levels of intelligence differed between races also played a role in the passage of the Immigration Act of 1924, which severely limited immigration into the United States from southern and eastern Europe until the mid-1960s. And in their 1994 book *The Bell Curve,* Richard Herrnstein and Charles Murray worried that society was becoming stratified by intelligence into a "cognitive elite" at the top and a growing underclass of the "custodial state."

Martin Kallikak's Children

In 1865, Francis Galton published a remarkable two-part article in *Macmillan's Magazine,* a popular Victorian periodical, titled "Hereditary Talent and Character."[1] His idea was to identify men of talent using appropriate compendia following which he would try to establish their pedigrees. Galton believed that if talent and character were inherited, then eminent male relatives should appear in these pedigrees in higher frequency than they did in the population as a whole. The results he obtained led him to argue that his thesis was supported even though he had to acknowledge that influence, for example of an eminent father in obtaining a desirable position for his son, might also play a role. In 1869, Galton expanded this approach greatly in his book *Hereditary Genius* where he obtained pedigrees of famous judges, statesmen, military commanders, and so forth.[2] He argued that if talent and character were inherited, then the closest male relatives to the eminent man (e.g., fathers and sons) were more likely to be eminent than more distant relatives (e.g., grandfathers and grandsons). He concluded that his results supported this expectation.

Galton had also become intrigued in the application of the normal distribution, particularly whether it might be somehow applied to data for mental ability. As a proxy, he used examination marks for admission to the Royal Military College at Sandhurst. The data clearly fit a normal distribution at the upper tail and in the center, but there were no data for those who got low scores. Galton also constructed a hypothetical normal distribution to classify British men "according to their natural gifts" while estimating the population of the United Kingdom to be about 15 million. The latter curve was eerily reminiscent of normal distributions of IQ that would be published much later on.

Although Galton collected and analyzed anthropometric data, investigated mental imagery, and used composite photography to see if criminals, soldiers, and so on shared unique facial features, he never discovered a method for measuring mental ability.[3] However, shortly before Galton's death, Alfred Binet did invent such a test.[4] Binet, a psychologist, had become interested in mentally handicapped children. Under new French laws mandating universal education, such children were required to attend school, whereas earlier they would have dropped out or never have attended school at all. Because they could not keep up with normal curricular requirements, they needed special attention. In 1904, Binet joined a government commission investigating the status of the mentally handicapped in France. The commission concluded that accurate diagnosis posed the biggest problem.

In 1905, Binet, together with Theodore Simon, a young physician who was studying psychology with him, began to try to devise a test that would unequivocally differentiate between normal and subnormal children. They avoided tests that relied heavily on reading and writing so as not to confuse intelligence with lack of schooling. Many of the items they chose assumed a familiarity with French life and were based on work that Binet had done earlier using his own two daughters as well as other children as experimental material. After some initial frustration, Binet and Simon realized that age had to be taken into account in differentiating between normal and subnormal children because both groups could learn to pass the same tests, but normal children did so at a younger age. The original tests had focused mainly on the most retarded and youngest children, so Binet and Simon set out to revise their intelligence tests in 1908 and again in 1911 because some of the most difficult educational decisions involved older children who were closer to being normal. Accordingly, Binet and Simon were eventually able to arrive at an age-standardized system that could compute each child's intellectual level.

Henry Herbert Goddard brought Binet's tests to the United States.[5] Goddard received his PhD in Psychology from Clark University and subsequently began to study children at the Vineland Training School for Feeble-Minded Girls and Boys in New Jersey, where he became director of the Psychological Research Laboratory, newly established in 1906 by its superintendent, Edward Johnstone.[6] In 1908, Goddard traveled to Europe seeking new techniques for determining mental ability. He had become disenchanted with the psychophysical measurements devised by James Cattell, based on the anthropometric methods of Francis Galton, as being of any help in classifying mental deficiency.[7]

While in Europe, Goddard learned about the Binet-Simon tests whose significance he soon appreciated. Upon his return to the United States, he began to apply the tests to children in the Training School.[8] Unlike Cattell's psychophysical measurements, the Binet-Simon tests did seem to classify children in a manner that was consistent with the observations made by his staff. The test results revealed the existence of wide variations in the degree of "feeblemindedness" and Goddard used the numbers he obtained to erect specific categories. Those children with a mental age of one or two were classified as "idiots," those with a mental age of three to seven were "imbeciles," whereas those whose mental ages were between eight and twelve were dubbed "morons," a term Goddard himself took from the Greek word for dull or stupid.

Up to this point, Goddard's work had no hereditarian implications, but in 1912, he published *The Kallikak Family: A Study in the Heredity of Feeblemindedness*. The name Kallikak was a pseudonym constructed from the Greek words kalós (good) and kakós (bad). Goddard had long suspected that feeblemindedness might have a heritable component as his earlier field studies revealed that some families seemed to have a high incidence of mental disease. The book followed the lives of a young Vineland woman, pseudonymously named Deborah Kallikak, and her relatives. The evidence, obtained by Goddard's fieldworker Elizabeth Kite, who had employed interviews rather than Binet-Simon tests, nevertheless seemed convincing to Goddard.

Deborah's great-great grandfather, Martin Kallikak Sr., a Revolutionary War soldier, had an affair with a feebleminded woman when he was a youth. She gave birth to Martin Kallikak Jr. and among 480 descendants, 143 were feebleminded with only 46 known to be normal, with the mental abilities of the rest unknown.

On leaving the Revolutionary Army, Martin Sr. "married a respectable girl of good family" and their 496 descendants were all normal. What could be clearer? Feeblemindedness was obviously inherited and probably Mendelian in nature. In fact, Goddard's results were bolstered by the many pedigrees presented principally by American and British workers at the First International Congress of Eugenics held in London in 1912. They appeared to show that the Mendelian inheritance of feeblemindedness and other undesirable traits (e.g., tuberculosis, "criminalistic tendencies," alcoholism) in individual pedigrees. Also, research conducted by fieldworkers of the Eugenics Record Office at Cold Spring Harbor, New York, directed by Charles Davenport, poured out a steady stream of supporting results. So, two powerful ideas now interacted to bolster the case for sterilization or segregation of the feebleminded and those with related disabilities. Numerous pedigrees alleged that these traits were inherited as Mendelian characters, whereas IQ tests could be used to identify individuals of subnormal intelligence irrespective of heredity.

Goddard's work encouraged others to experiment with methods designed to assess the quantitative properties of mental ability. One of these was Lewis Terman, a Stanford University psychologist, who published a revision of the Binet-Simon test that thereafter became known as the Stanford-Binet test.[9] Terman, a hereditarian like Goddard, introduced the term IQ. This, of course, stood for intelligence quotient, a number invented by the German psychologist William Stern in 1912 in which he multiplied the ratio of a child's mental age to his actual age by 100. Hence, an average child would have a ratio of 1 or an IQ of 100.

Meanwhile, Charles Spearman, a British psychologist and a fine statistician, invented a totally new concept of general intelligence that he represented by the symbol *g*.[10] What Spearman found was that if two mental tests were given to a large number of individuals, the correlation coefficient between them was nearly always positive. Spearman realized that one explanation of his results was that these positive correlations might be reduced to a single underlying factor, *g*. To distinguish whether a single factor or multiple factors explained the positive correlations, Spearman invented a new statistical technique, factor analysis. He concluded that a single underlying factor was indeed involved and published his theory in 1904. Spearman's concept of g seems to have withstood the test of time pretty well.

As it became increasingly clear that the United States was about to enter World War I, the National Academy of Sciences established the National Research Council to mobilize scientists for defense.[11] In May 1917, a group of psychologists was organized with the blessing of the council to design a testing system that would aid the army in placing its men in proper categories with respect to military occupation. Goddard and Terman were included, but the chief tester was Robert Yerkes, a young Harvard faculty member, who together with James Bridges, a Harvard graduate student in psychology, developed the Yerkes-Bridges scale, a rival to Terman's Stanford-Binet test. These psychologists proceeded to improvise two intelligence tests called the Army Alpha and the Army Beta. The Army Alpha consisted of problems and questions that have become familiar to many generations of students: word analogies, sequences of numbers, arithmetic problems, synonym-antonym puzzles, and queries that required commonsense answers. The Army Beta tests were pictorial and given to illiterates and those who failed the Army Alpha tests. Eventually, 1.75 million men were tested and Yerkes wrote a huge monograph titled *Psychological Examining in the United States Army* (1921).

The major conclusions were that the average mental age of white American men was just above that of a moron; that European immigrants could be graded by country of origin with southern and eastern Europeans being less intelligent than those from northern and western Europe; and blacks were at the bottom. So, the tests conveniently reinforced existing prejudices except for the embarrassment concerning the average white male. In 1923, another of the army testers, Carl Brigham, an enthusiastic Yerkes disciple, published *A Study of American Intelligence*.[12] He reported that the army tests revealed the first significant differences in mental abilities between different races, claiming that the average intelligence of black Americans was low compared with white Americans, thus opening a debate that continues to this day. He also claimed that the Alpine and Mediterranean "races" were "intellectually inferior to the Nordic race" and that the average intelligence of immigrants was declining. This kind of "evidence" would be used in support of the Immigration Act of 1924, which restricted the influx of immigrants from southern and eastern Europe.

Six years later, Brigham recanted having decided that the army tests failed to measure innate intelligence accurately.[13] Goddard also realized that these tests must be flawed as it made little sense that the average mental age of white recruits was just above that of a moron, the category he had created for the feebleminded who performed best on IQ tests. In 1928, he published an article titled "Feeblemindedness: A Question of Definition" in which he concluded that feeblemindedness was not incurable and that the "feeble-minded do not generally need to be segregated in institutions."[14] In Goddard's own words, he had "gone over to the enemy." The last of the hereditarian psychologists to admit defeat did so in a much more subtle way. In 1937, Terman revised his 1916 book on the measurement of intelligence.[15] He no longer argued that differences between groups in IQ could be explained by heredity, but now framed them in environmental terms.

At the same time, it was becoming clear from the investigations of Lionel Penrose at the Royal Eastern Counties' Institution in Colchester, England (refer to Chapter 2, "How genetic diseases arise"), that "feeblemindedness" was actually a composite description that included, but was not circumscribed, by discreet genetic diseases such as phenylketonuria and Down syndrome. A vocal critic of the hereditarian interpretation of intelligence throughout was the young journalist Walter Lippman. He was adept at identifying the flaws in the hereditarian arguments, pointing out that intelligence "is not an abstraction like length and weight; it is an exceedingly complicated notion which nobody has as yet succeeded in defining."[16]

But before we move on, let us return to Carl Brigham for a moment. After assisting Robert Yerkes in designing the Army Alpha tests, Brigham joined the faculty of Princeton University. In 1925, the College Board asked Brigham to try to design a test for college admission that could be applied universally.[17] To meet this requirement, Brigham tinkered with the Army Alpha, principally by making it harder. In 1926, he administered his test experimentally to a few thousand college applicants. By 1933, he had concluded that his scholastic aptitude test (SAT) reliably predicted the likelihood of academic success.

Meanwhile in 1933, James Bryant Conant was appointed President of Harvard University. He wanted to increase the breadth of Harvard's student body beyond the eastern preparatory schools that supplied it. His solution was to initiate a scholarship program for gifted boys from elsewhere. But how were applicants to the new program going to be evaluated? Conant delegated the problem of finding a solution to Henry Chauncey, an assistant dean. Chauncey met Brigham and, impressed with his test, recommended that Harvard adopt the test.

But then Brigham began to have doubts about his test, a development that Conant, and especially Chauncey, found irritating for by 1938, Chauncey had convinced all the member schools of the College Board that they should use the test, but only for scholarship applicants. Brigham's opposition prevented the test from being generally adopted, but his premature death at the age of 52 in 1943 cleared the way for acceptance of the SAT. It became the standard test for all applicants in 1944. In that year, under contract to the army and the navy, Chauncey administered the SAT to 300,000 officer candidates from all over the United States in a single day. In 1947, the Educational Testing Service was chartered as a nonprofit organization with Chauncey as its first president. Today, millions of college applicants take the SAT every year. Over the years, the test has changed in some respects, but not completely.

The rise of behaviorism

During the very period when the hereditary determinist view of intelligence was at its peak, a new theory that was its antithesis, behaviorism, was emerging. The main proponent of what came to be called classic behaviorism was John Broadus Watson.[18] Watson, a professor at Johns Hopkins University, coined the term in a 1913 article in *Psychological Review* titled "Psychology as the Behaviorist Views It."[19] In that paper, Watson stated that psychology "as the behaviorist views it is a purely objective branch of experimental science. Introspection forms no essential part of its methods, nor is the scientific value of its data dependent upon the readiness with which they lend themselves to interpretation in terms of consciousness."

Watson here was stating his belief that psychology, as he envisioned it, was a hard science in which inferences about one's own consciousness or the inferred consciousness of anyone else had no place. Watson went on to say that there was "no dividing line between man and brute." He eschewed the anthropocentric approach be it in terms of consciousness or as in Darwin's time when "the whole Darwinian movement was judged by the bearing it had upon the origin and development of the human race." But the "moment zoology undertook the experimental study of evolution and descent, the situation immediately changed. Man ceased to be the center of reference."

Later, Watson went on to define more precisely what he meant when he said that the sort of psychology that he wanted "to build up would take as a starting point, first, the observable fact that organisms, man and animal alike, do adjust themselves to their environment by means of hereditary and habit equipments." But despite his acknowledgment of the importance of "hereditary equipment" in determining behavior, it was clear that Watson's goal was to control behavior in a predictable way so that "given the response the stimuli can be predicted; given the stimuli the response can be predicted." One could achieve this end by placing an "animal in a situation where he will respond (or learn to respond)" to a specific stimulus.

The key phrase here of course is "learn to respond" and in 1920, Watson published a classic paper in the *Journal of Experimental Psychology* with his graduate student Rosalie Rayner titled "Conditioned Emotional Reactions."[20] The subject of this paper was an infant named Albert B who had been reared in a hospital where his mother was a wet nurse. He was stolid, unemotional, and rarely cried. Watson and Rayner decided to see whether a fear reaction could be induced in "Little Albert," although with "considerable hesitation" as "a certain responsibility attaches to such a procedure." When Albert was almost 9 months old, Watson and Rayner attempted to induce the fear response by hitting a steel bar with a hammer behind the child's back. The child "started violently" the first two times the bar was hit and the third time began to cry. At a little over 11 months, Albert was presented with a white rat that he reached for. "Just as his hand touched the animal the bar was struck immediately behind his head." Albert "jumped violently and fell forward." The experiments continued in this vein and were expanded to include a rabbit, a dog, and a sealskin coat. The upshot was that Albert became afraid of furry things. Although there was "a certain loss of intensity with time," the investigators concluded these "conditioned emotional responses" probably "persist and modify personality throughout life."

It seems that Watson and Rayner, his graduate student, also had a conditioned emotional response to each other. Watson's marriage to Mary Ickes, the sister of Harold Ickes, later Roosevelt's Secretary of the Interior, had been under strain for many years. Nevertheless, Watson's liaison with Rosalie Rayner and the subsequent divorce proceedings caused a national sensation. As a consequence of his adultery with his student, Watson was fired from his professorial post at Johns Hopkins in 1920. Through a friend of his, Watson landed a job at J. Walter Thompson, the advertising agency, where he applied behavioral psychology promoting Pond's Facial Cream, for example, with considerable success.

Watson continued to publish actively and in his 1924 book *Behaviorism,* coined one of the best-known statements in psychology.

> Give me a dozen healthy infants, well-formed and my own specified world to bring them up in and I'll guarantee to take any one at random and train him to become any type of specialist I might select—doctor, lawyer, artist, merchant-chief, and, yes, even beggarman and thief, regardless of his talents, penchants, tendencies, abilities, vocations, and race of his ancestors. I am going beyond my facts and I admit it, but so have the advocates of the contrary and they have been doing it for many thousands of years.[21]

Watson's behaviorism served as the impetus that propelled Burrhus Frederic Skinner into the field.[22] Following his graduation from Hamilton College, Skinner returned to his parents' home in Scranton, Pennsylvania, built a study in their attic, and attempted, without notable success, to write fiction. But Skinner read widely and one of his favorite authors was Bertrand Russell. In one of his books, Russell presented a critique of Watson's *Behaviorism.* Later, Skinner read an article by H. G. Wells about Ivan Pavlov and George Bernard Shaw. Shaw detested Pavlov for he felt he was a vivisectionist harming dogs through his operations just to see what might happen.

But Skinner was intrigued by Pavlov's experiments as he had been by Watson's behaviorism and decided to make his career in behavioral psychology. He was accepted as a graduate student at Harvard progressing through a postdoctoral fellowship to be invited to become a junior fellow in Harvard's distinguished Society of Fellows. During his eight years at Harvard, he had much opportunity to experiment and the culmination of these experiments was the Skinner box. When a white rat was introduced into this box, it discovered, after suitable exploration, that by pressing a bar it could release a food pellet into a tray. The rat soon caught on as shown by the cumulative record Skinner was able to keep. Skinner called the provisioning of food in response to pressing the bar "reinforcement" and named this kind of learned behavior "operant conditioning." Skinner had faculty positions at the University of Minnesota and Indiana University before returning to Harvard in 1948.

Meanwhile, he continued to improve and refine the notion of reinforcement. Reinforcers, like the provision of food when the rat pressed on a bar, were positive in nature, but there was negative reinforcement too. This means strengthening a particular behavior by avoiding or stopping a condition perceived as being negative. Driving in heavy traffic is unpleasant for most of us. Suppose you go to work earlier than you planned and you encounter less traffic. If you decide that you will continue to drive to your place of employment early to avoid heavy traffic, you are avoiding a condition you find unpleasant (negative reinforcement).

Meanwhile, Skinner was collecting students and postdoctoral fellows and becoming famous. In 1948, he published *Walden Two,* a utopian novel describing an ideal society in which negative reinforcement was nonexistent and children sought only positive reinforcement.[23] Consequently, they grew up to be happy, sociable, and civilized. Frazier, the hero of the novel, is accused by a skeptical visitor named Castle of being a despot because Frazier believes that through the use of positive reinforcement, a planned society will be created. Frazier remarks, "it's a limited sort of despotism" that should not worry anyone because the "despot must wield his power for the good of others." If he "reduces the sum total of human happiness his power is reduced by a like amount. What better check against a malevolent despotism could you ask for?"

As might be expected, *Walden Two* proved controversial. Critics felt Skinner's happy, but controlled society was in fact totalitarian, whereas others responded positively. In 1971, Skinner produced a best-selling book titled *Beyond Freedom and Dignity* in which he argued that the notion of "autonomous man," which is the foundation of many of the institutions of Western society, is misguided.[24] Although we usually "credit" people with doing good deeds "of their own free will" rather than because they are required to, in reality we simply do not know the contingencies that resulted in the behavior in the first instance whereas we do in the second. In the end, behavioral freedom is an illusory concept and the reality is that behavior is conditioned most effectively over the long term by positive as opposed to negative reinforcement. Once this principle is accepted, it becomes incumbent for societies to become like the one described in *Walden*

Two where behavior is shaped to socially desirable ends through positive reinforcement. In Skinner's world, behavioral conditioning and not genes determined who and what we are and it was this viewpoint that largely prevailed for some time after World War II. In fact, Skinner was listed in the *The 100 Most Important People in the World* and as the best-known scientist in America in a 1975 survey.

The return of the hereditarians

By the late 1930s, it had become clear that pedigrees were an inappropriate way to study mental ability except in the cases of well-defined genetic diseases like phenylketoneuria that Penrose had shown to affect the capacity of the brain to function properly. However, studies of IQ measurements in foster children, adopted children, and twins were starting to be reported.

As it turns out, Francis Galton was the first to use comparisons of twins to assess the roles of nature and nurture in determining character. Identical (monozygotic, MZ) twins are genetically indistinguishable, whereas nonidentical or fraternal (dizygotic, DZ) twins are no more alike genetically than brothers or sisters. The first derive from a single fertilized egg, whereas the latter arise from two independently fertilized eggs. Galton clearly recognized this distinction and it was critical to his analysis. MZ twins did not differ either in nature or nurture, whereas DZ twins were reared in the same environment, but differed genetically.

Armed with a psychological questionnaire he had invented, Galton eventually networked his way to 94 pairs of twins, but only reported on 35 of these pairs. He classified his responses by sex and by whether the twins were "alike," "unlike," or "partially alike." He published his findings in 1875 in *Fraser's Magazine,* a popular Victorian periodical.[25] In the course of his work, Galton collected anecdotes about the twins and recounted the more remarkable of these. Some dealt with the strikingly close physical similarity of certain twins, whereas others emphasized shared behavioral quirks. Psychologists have made note of such quirks ever since.[26] Although it is quite likely that some, or perhaps many, of the behavioral quirks described have a genetic basis, they add to the overall impression that intelligence has a strong heritable component, even though they themselves may not relate to intelligence as measured by IQ.

Frank N. Freeman and Karl J. Holzinger, two University of Chicago psychologists, in collaboration with Horatio H. Newman, a biologist, carried out the most important early study of intelligence in identical twins.[27] In 1937, they reported on the IQs of 19 pairs of twins reared apart compared with a similar number reared together. The correlation of 0.67 for the twins reared apart implied a strong hereditary component. However, they also noted that the greatest IQ differences were found for those twins reared in the most strikingly different environments. The three scientists were cautious in their conclusions saying that both nature and nurture seemed to play important roles in producing the correlation. This study, supported by other findings, led to a consensus that both heredity and environment were important and interacting contributors to human intelligence. This comfortable situation remained until the famous British psychologist Sir Cyril Burt reported intelligence test correlations of approximately .771 for ever-increasing numbers of pairs of identical twins reared apart in three papers published in 1943, 1955, and 1966, with the latter paper including 53 pairs.[28] The correlations had a suspicious consistency.

Meanwhile, the Berkeley psychologist Arthur Jensen had been invited to write a major article on the question "How much can we boost IQ and scholastic achievement?" by the student editors of *Harvard Educational Review.*[29] His conclusion, stated at the beginning of the article, was that compensatory education had failed and that large programs designed to improve the scholastic aptitudes of culturally deprived children either were unsuccessful or resulted in minor gains at best. Jensen went on to state that nature was more important than nurture in determining intelligence and cited a number of studies supporting this conclusion, but most notably those of Burt. He concluded that IQ heritability ranged from .70 to .90 centering near .80.

Jensen's article was long and scholarly, but he got into deep trouble in a short section of the paper that compared the IQs of blacks and whites. There, Jensen noted that there was a vast amount of literature suggesting that on average the IQ of blacks was 15 points below the IQ of whites. When gross socioeconomic differences were accounted for, this number reduced to 11 points that was "about the same spread as the average difference between siblings in the same family." But a little later, Jensen wrote that it seemed "not unreasonable, in view of the fact that intelligence variation has a large genetic component, to hypothesize that genetic factors might play a part in this picture." He pointed out that the notion that black/white IQ differences might in part be hereditary in origin had been strongly denounced, but "it has been neither contradicted nor discredited by the evidence." And then he took a stand. "The preponderance of the evidence is, in my opinion, less consistent with a strictly environmental hypothesis than with a genetic hypothesis, which, of course, does not exclude the influence of environment or its interaction with genetic factors." With those words, Jensen had let an evil genie out of its bottle.

The student editors of the *Harvard Educational Review* were so taken aback with the negative response Jensen's article had received that they denied having ever solicited it even though he sent them a copy of his original letter of invitation.[30] Feature articles in *TIME, Newsweek, LIFE, U.S. News & World Report,* the *New York Times Magazine,* and many other newspapers and magazines, as well as radio and TV programs jumped on Jensen's claims. Jensen became a pariah on his own campus and subject to much abuse. Despite all of this, Jensen was not deterred, but has continued to publish a stream of books and articles on mental tests and what they mean. In 1982, he also proclaimed the demise of Watson's behaviorism, arguing that IQ is much less plastic than the behaviorists supposed as genetic variance "in IQ and information-processing capacity are strongly influenced by hereditary factors, with genetic variance constituting about 70% of the total population variance in IQ."[31] That would leave the behaviorists with only 30% of the variance to work with.

In 1971, Harvard psychology professor Richard Herrnstein got himself into hot water in a popular article about IQ testing in the *Atlantic* that he hoped would promote debate.[32] After reviewing the history of IQ testing, Herrnstein discussed the advantages of having a high IQ, one of which was economic. Salaries of those with high IQs were greater than those with low IQs. Herrnstein then averred that IQ appeared to have a large heritable component, echoing Jensen's contention that it was not possible to boost IQ and scholastic achievement very much. Herrnstein considered IQ differences between blacks and whites, pointing out that Jensen's data concerning the heritability of IQ had been based almost entirely on whites, a point Jensen himself acknowledged. Herrnstein continued that although "there are scraps of evidence for a genetic component in the black/white difference, the overwhelming case is for believing that American blacks have been at an environmental disadvantage." But, he continued, setting "aside the racial issue, the conclusion about intelligence is that, like other important though not necessarily vital traits, it is highly heritable."

Toward the end of his article, Herrnstein imagined a world in which "the social classes not only continue but become ever more solidly built on inborn differences." He envisioned that as "the wealth and complexity of human society" grew "there will be precipitated out of the mass of humanity a low-capacity (intellectual and otherwise) residue that may be unable to master the common occupations, cannot compete for success and achievement, and are most likely to be born to parents who have similarly failed." Herrnstein, in short, was beginning to develop themes that would emerge over two decades later in *The Bell Curve.* Fellow scientists, pro and con, were generally constrained in their responses, but student radicals pilloried Herrnstein much as they had Jensen.

The scientific counterattack on the hereditarian view of IQ, now defined in the dictionary as "Jensenism," began with a discovery made by Princeton psychologist Leon Kamin. After reading Herrnstein's article, Kamin realized that Burt's twin studies were the foundation upon which the hereditarian thesis rested. So Kamin set to work to examine Burt's results and made a surprising discovery.[33] In what Burt described as the "Group Test," Burt's correlations were constant at .771 in his 1955 paper on 21 twin pairs reared apart, in "over thirty" pairs reported in 1958, and in the 53 pairs discussed in his 1966 paper. Subsequently, Kamin found other instances of invariant correlations, leading him to believe that Burt's twin studies had been faked.

Kamin then examined other twin studies and found these correlations variable. Moreover, he showed that twins not genuinely separated in these studies were more alike in IQ than those who really were reared separately where bigger differences emerged. Kamin was similarly critical of Jensen's adoption studies. Meanwhile, the scandal around Burt, who had died in 1971, mushroomed and has seesawed back and forth ever since, although it now seems pretty clear that the twins data were fraudulent.[34] Soon thereafter, other opponents weighed in on the controversy, notably Harvard population geneticist Richard Lewontin and his colleague, paleontologist Stephen Jay Gould whose popular book *The Mismeasure of Man* (1981) was aimed at debunking the hereditarian theory of intelligence.[35] Several years later, Lewontin, in collaboration with Leon Kamin and Steven Rose, published a book titled *Not In Our Genes* (1984).[36] In that book, they not only questioned what IQ was really measuring, but also raised serious concerns once again about adoption studies and those of identical twins reared apart.

For a while, the hereditarian thesis seemed to go into decline, but it reemerged with a vengeance with the publication of *The Bell Curve* in 1994 by Richard Herrnstein and Charles Murray[37] together with a companion article in the *New Republic* titled "Race, Genes, and I.Q.—An Apologia."[38] The book, divided into four parts, made the case that intelligence (as measured by IQ) rather than socioeconomic status is the major determinant of success in life; that twin studies showed that IQ has a major heritable component; and that there was a 15-point average IQ difference between blacks and whites. This latter point was brought out most starkly in the *New Republic* article. Because of all of this, Herrnstein and Murray argued that a "cognitive elite" was evolving and that society was becoming stratified because of an emerging economic system that rewards intelligence above all else.

The last section of the book dealt with policy matters. There, Herrnstein and Murray asked the question, "Can people become smarter if they are given the right kind of help?" They answered negatively, writing that cognitive gains achieved by children in projects like Head Start vanished by the end of primary school. Herrnstein and Murray envisioned a future America dividing into an affluent aristocracy of the "cognitive elite" while at the bottom would be the growing underclass of the "custodial state." The masses in the latter category would be sternly controlled by the welfare state "a high-tech and more lavish version of the Indian reservation." To avoid this outcome, Herrnstein and Murray argued that current efforts to achieve "equality of outcomes" should be abandoned. People should be allowed to find their own "valued places in society" according to their ability. In a "simpler America," even people of lower intelligence were valued and nurtured in the local community. Federal centralization has "drained" life out the community and more social functions should be returned to the neighborhood. Also, regulations relating to criminal justice, small business, and so forth needed to be simplified so the less bright could understand them.

The Bell Curve, a hardcover book of over 800 pages, sold over 500,000 copies and was reviewed on countless occasions in the public and academic press. Much of the commentary was negative, ridiculing the hereditarian bias that the authors introduced into their arguments, especially where race was concerned. *The Bell Curve* also became the focus of entire books bent on rebuttal. Claude S. Fischer and colleagues of his in the Department of Sociology at the University of California, Berkeley, made a particularly perceptive analysis of the *The Bell Curve.*[39] Not only did they uncover basic technical errors in the book, but they found numerous cases where Herrnstein and Murray had presented data seeming to document that intelligence was of greater importance than socioeconomic status in determining success where the reverse was, in fact, true. For example, Herrnstein and Murray had claimed that the probability of being poor was more a function of intelligence level than socioeconomic status, but Fischer et al. found that socioeconomic status was actually more important.

They also demonstrated that the questions on the Armed Forces Qualifying Test (AFQT) used by Herrnstein and Murray as their measure of IQ were geared to the school achievement level of the individuals being examined. Thus, a student who had done well in school was likely to score higher on the AFQT than a poor student who might have dropped out of high school along the way. They also argued that the reason blacks and Latinos average lower than whites on tests like the AFQT had to do more with socioeconomic deprivation, including poor schools, than it did with heredity. Perhaps most important of all, Fischer and colleagues seriously questioned whether psychometric methods using IQ tests as tools were really useful in measuring intelligence at all. They pointed to Arthur Jensen's quip that "intelligence is what intelligence tests measure."

Balancing nature with nurture

In *The Bell Curve*, Herrnstein and Murray discussed a phenomenon that they called "The Flynn Effect" as a possible environmental explanation for the 15-point mean difference between the IQs of whites and blacks.[40] They picked this name because James R. Flynn, a professor of political science at the University of Otago in Dunedin, New Zealand, made the remarkable discovery that during the twentieth century, there were massive gains in IQ from generation to generation.[41] Furthermore, data obtained from a Dutch colleague showed that in young Dutch males, these gains could occur within a single generation. The gains exhibited linearity over time and were greater in certain subtests (similarities) and much less dramatic in others (information & arithmetic & vocabulary). Because Flynn found striking IQ gains not only between generations, but also within a generation, these gains could not be explained on the basis of heredity. Furthermore, the differences in IQ increases between different subtests showed that the overall test scores reflected an internal heterogeneity.

Then, Nicholas J. Mackintosh, a distinguished animal-learning theorist at the University of Cambridge, decided to wade into the controversy over nature, nurture, and intelligence.[42] He came into this fraught field without any particular prejudices and proceeded to sort through much of the relevant literature coming up with a balanced view on the subject. Mackintosh summarized the data on MZ twins reared apart. By now, this included the work done by Thomas Bouchard, the Director of the Minnesota Center for Twin and Adoption Research at the University of Minnesota, and his colleagues. Bouchard's studies were the most recent and complete studies reported to date. Macintosh also included the four other similar studies published earlier, omitting Burt's papers because, if not fraudulent, they were laced with errors.

Macintosh observed that these reports gave a mean correlation in IQ between MZ twins reared by different parents that range from 0.69 to 0.78. By contrast, the correlation between the IQ scores of adopted children and those of their adoptive parents was usually less than between parents and their biological children. Furthermore, the IQs of these adopted children generally resembled those of their biological mothers more closely than they did their adoptive mothers.

At first sight, these findings seemed to indicate that IQ was a highly heritable trait, but closer scrutiny suggested that it was dangerous to take these numbers too literally because the sum total of MZ twins examined in these five studies was only 162. Furthermore, it became apparent to Macintosh that the home environments in which the twins were brought up were not wholly uncorrelated. For example, there were significant correlations between the father's occupation, the mother's education, material possessions at home, as well as family background. However, some of these variables (e.g., father's occupation and material possessions at home) were themselves correlated so they could not independently contribute to the IQ similarities between the separated twins.

Mackintosh concluded that some of the assumptions of what he called the "simple model" for interpreting the results of twin and adoption studies were false. For example, the model assumed that all members of a family share a uniform environment. "But MZ twins may well share more experiences in common than DZ twins, who in turn may share more common experiences than siblings of different ages." With regard to adoption studies, no provision was made for the likelihood that adoptive families were not a representative cross-section of the population, or for selective placement of adoptive children. Mackintosh prudently concluded that "the broad heritability of IQ in modern industrialized societies is probably somewhere between 0.30 and 0.75, and that neither the data nor the models justify much greater precision."

Mackintosh also addressed the much-reported average IQ difference between blacks and whites of 15 points. As Herrnstein and Murray took pains to point out in their 1994 article, "Race, Genes and I.Q.—An Apologia," this difference, which amounts to 1.2 standard deviations, "means considerable overlap in the cognitive ability distribution for blacks and whites."[43] Mackintosh analyzed the relevant data in detail concluding that, "there is remarkably little evidence that the difference is genetic in origin." But if "the effect *is* an environmental one, it remains to be shown what the proximal environmental causes are." The same problems arose in trying to sort out why Chinese and Japanese seem to score higher than whites on spatial IQ tests and possibly on other nonverbal tests. Then, there was the question of whether IQ tests were in some way biased against certain ethnic groups. Mackintosh argued that the only way to tell whether an IQ test is biased is to compare the results with "an independent, better measure of 'intelligence.'" But there was no such test.

Richard Nisbett has echoed many of the concerns raised by Macintosh.[44] Nisbett is a social psychologist and distinguished university professor at the University of Michigan and the first psychologist in a generation to be elected to the National Academy of Sciences.[45] Nisbett pointed out that individuals like Arthur Jensen, whom he referred to as "strong hereditarians," believe that the heritability of IQ is 75% to 85%, whereas "the environmentalist camp estimates heritability to be 50% or less." He proceeded to cast serious doubt on the hereditarian position, pointing out, much as Macintosh had before him, that for MZ twins reared apart "a tacit assumption that is surely false" is that "the twins were placed into environments at random." Nisbett mentioned, for example, the work of Urie Bronfenbrenner who showed that, when twins reared apart were brought up in very similar environments, the correlation between their IQ scores was .83 to .91. However, when environments were not alike, the correlations ranged from .26 to .67.

Eric Turkheimer at the University of Virginia took a different approach. Turkheimer studied over 600 pairs of twins reared together most of whom were in families below the poverty level.[46] He "found that for the poorest twins, IQ seemed to be determined almost exclusively by their socioeconomic status, which is to say their impoverished environment. Yet, for the best-off families, genes were the most important factor to determining IQ, with environment playing a much less important role." Turkheimer's findings suggested that upper-middle-class families were providing a much better environment for the development of intelligence and they may not differ a great deal between themselves in this regard. Because studies of twins reared apart tend to be conducted using middle-class and upper-middle-class families who are relatively easy to contact, this will bias estimates of heritability toward the high end.

Nisbett also weighed in on the factors responsible for the 15-point average differential in IQ scores between blacks and whites debunking heredity as being responsible. For one thing, an interesting natural experiment occurred after World War II when American

GIs fathered children by German women. Some of these soldiers were black and some were white. German children of white fathers averaged IQs of 97.0, whereas those of black fathers averaged 96.5. Nisbett was careful to point out that American blacks have an average of 20% European genes, so they ranged in genetic composition from almost purely African to mostly European.

Then there was the work of psychologists Joseph Fagan and Cynthia Holland. They tested black and white community college students on their ability to learn and reason using words and concepts. Although whites had a broader knowledge of words and concepts, the two sets of participants had an equal ability to learn new words, for example, from a dictionary. Whites also comprehended sayings better than blacks, could recognize similarities better, and had greater facility with analogies whose solutions depended upon a knowledge of words and concepts that were more likely to be more familiar to whites. But when this sort of reasoning was restricted to words and concepts equally familiar to blacks and whites, no differences were apparent.

Psychologists like Mackintosh and Nisbett have seriously challenged the hereditarian thesis. They argue that intelligence is malleable rather than fixed. You can improve on your intelligence through learning. Certainly, most would-be college or professional school students would agree. They usually take advantage of the test preparation services offered by Kaplan, Inc., founded in 1938 by Stanley Kaplan.[47] The College Entrance Examination Board, which oversaw the SAT believed it was uncoachable.[48] Mr. Kaplan thought otherwise. He regarded himself at the nemesis of the Educational Testing Service that develops the SAT and other tests. Kaplan claimed he could improve SAT scores by 100 points, but a Federal Trade Commission investigation of the coaching business revealed that Kaplan's program helped participants raise their scores by only 50 points. That is no mean achievement and it has nothing to do with genes. It would be interesting to know what Carl Brigham, who developed the original SAT, would have thought of all this.

Stalking intelligence genes

As this brief history shows, question of the genetic contribution to IQ and, hence, to intelligence has been the subject of debate for over a century. Opinion has seesawed back and forth from the notion that genes are the principal determinants of IQ to the belief that the environment mainly determines IQ and back again. The principal problem is that the available experimental approaches have evolved very little over time. Since 1937, when Freeman and their colleagues published their original study, we have been largely dependent on studies of MZ twins reared apart for estimates of the heritability of IQ. A more precise understanding of the genetic contribution to intelligence is going to require a better idea of the spectrum of genes involved and their roles and acquiring this knowledge will take time.

One of the reasons that twins have been relied on so heavily is that IQ is a complex trait. Oddly enough, it is similar to height in several respects. Like height, IQ is a metric trait so the distribution of IQ scores is easily plotted and, like height again, a normal distribution results. Traits like these are referred to as being polygenic because many genes, usually having small effects individually, are involved in determining the distribution. Sorting these out is going to take time especially because the environment can also affect the distribution. For example, poor nutrition results in shorter height and lower IQ on average. Despite these drawbacks, characterization of genes affecting IQ should eventually provide a far more precise idea of the role that nature plays in determining intelligence.

So what do we know about genes affecting intelligence at the moment? One way of approaching this question is to ask how many genes possess mutations that cause mental retardation and what do the normal, nonmutated forms of these genes do? Obviously, some of these genes will affect neurological processes we associate with intelligence, but in other cases the effect will be indirect. For example, mental retardation in the case of phenylketoneuria results because the amino acid phenylalanine accumulates in the blood and in the brain. As of September 2003, 282 genes had been identified with mutations which caused mental retardation.[49] The majority of the genes identified could be classified as being involved in signaling pathways, metabolic pathways, and transcription, but numerous genes involved in neuronal or glial (nonneuronal cells that support neuronal activities) function were identified too.

Candidate gene and genome-wide association studies of intelligence are also under way.[50] More than 20 candidate genes have so far been identified, but many apparently significant associations have failed to replicate. Genome-wide association studies have identified single nucleotide polymorphisms (SNPs) and individual genes that seem to relate to intelligence, but none of these studies has been replicated well.

Future progress in understanding how genes contribute to intelligence is likely to come from collaborative investigations like those being carried out by the Genes to Cognition Project (G2C).[51] Investigators involved in this project are focusing on a complex structure called the NMDA receptor that seems to be involved in memory function and contains more than 100 proteins. NMDA receptors sit at the junctions (synapses) between nerve cells and modulate the strength of signals passing between neurons. Systematic investigation of the functions of the different proteins in this complex can be accomplished in mice where the genes encoding proteins of the complex can be knocked out one by one and the effect on learning assessed. For example, mice lacking a protein called PSD-95 that binds to the NMDA receptor have severe learning difficulties.

Genomic sequencing is becoming ever faster and cheaper. George Church at Harvard Medical School runs the Personal Genome Project that now has 16,000 volunteers.[52] Its goal is to sequence all of their genomes and eventually to expand the number to 100,000 genomes. So far, the genomes of 12 people have been sequenced. The sequences have been made publicly available. If Dr. Church, and others like him, succeed in sequencing so many genomes, it is fair to say that we will know a lot more about the genes involved in intelligence before too long. Understanding how their products interact in the brain and nervous system and how they interact with each other and with the environment may take much longer.

8

Preventing genetic disease

On June 26, 1974, just prior to her second birthday, Melisa Howard died of Tay-Sachs disease.[1] Her mother, Laura Howard, had been a patient of Dr. B. Douglass Lecher, a specialist in the field of obstetrics and gynecology. He had cared for Mrs. Howard during her previous pregnancy, performed regular gynecological examinations on Mrs. Howard prior to Melisa's birth, and was present when the baby was born.

Following Melisa's death, Mr. and Mrs. Howard, who were Jewish, sued Dr. Lecher claiming that his negligence had prevented them from terminating the pregnancy by abortion. They alleged that Dr. Lecher should have known that, because of their genetic background, the plaintiffs were "potential carriers of Tay-Sachs disease," and that tests were available that would have permitted diagnosis of Tay-Sachs disease in the fetus so it could have been aborted.

The Howards complaint sought damages for "severe mental distress and emotional disturbances resulting from the illness, and physical deterioration and death of their infant daughter and from the knowledge that no treatment or cure was available and that [she] would soon die." The Howards also sought recovery of "medical, hospital, nursing and related care and funeral expenses."

The trial court ruled that the Howards were entitled to seek damages for mental and emotional distress and that they were entitled to recover all their medical and related expenses. However, their request to be compensated for mental anguish was denied by the New York Court of Appeals. Four reasons were given for this decision: Lecher's failure to recommend the genetic test did not directly affect the parents themselves; the damages were impossible to calculate; recognition of the claim would be "unwarranted and dangerous extension of malpractice liability"; and were the claim sustained it would "either open the way for fraudulent claims or enter a field that has no sensible stopping point."

Two years after the Howard's filed their complaint, on December 27, 1978, the same New York Court of Appeals consolidated two cases in which the plaintiffs had also sued their doctors after they had given birth to genetically defective babies.[2] The first case involved Dolores Becker. In 1974, she became pregnant at the age of 37 and subsequently gave birth to a child with Down syndrome. Mrs. Becker stated that, had she realized that the genetic defect could have been detected in the embryo following amniocentesis, she would have opted for the procedure. The plaintiffs in the second case were Hetty and Steven Park. In 1969, she had a baby with polycystic kidney disease, a rare recessive genetic condition in children. The baby died within five hours. The Parks consulted their obstetricians about having another child and were incorrectly informed that the disease was not hereditary. Their second child was also born with polycystic kidney disease and died at the age of two and a half. The Court of Appeals consolidated the two cases and upheld judgments for the parents' claims in both cases. However, the court denied the child's claim of wrongful birth in both instances, hence establishing that under New York law a child cannot sue for being born with genetic defects.

But there was a quite different outcome in a California case.[3] Hyam and Phillis Curlender knew they were at risk of giving birth to a Tay-Sachs baby, so in 1977 they retained Bio-Sciences Laboratories and Automated Laboratory Services to perform blood tests on them to see whether they were likely to transmit Tay-Sachs genes to their offspring. DNA tests were not available at the time, but carriers could be distinguished from those lacking the mutant gene by assaying the enzyme hexosaminidase A. You will recall (refer to Chapter 3, "Ethnicity and genetic disease") that this enzyme consists of two dissimilar subunits and it is the α subunit that is rendered nonfunctional in Tay-Sachs disease. Therefore, carriers exhibit lower activity of this enzyme than those without the mutation.

Because the tests were reported to be negative, the Curlenders decided to have a baby, Mrs. Curlender conceived, but their daughter, Shauna, was diagnosed with Tay-Sachs disease. The Curlenders sued the testing laboratories, but in this case, the plaintiff was their stricken daughter Shauna. Their complaint sought costs of the plaintiff's care and also damages for emotional distress and deprivation of "72.6 years of her life." In addition, punitive damages of $3 million were requested because the plaintiff claimed that the defendant's scientific procedures were defective. Although the court denied the claim concerning deprivation of normal life expectancy, it ruled that Shauna had the right "to recover damages for pain and suffering during the limited lifespan available...and any pecuniary loss resulting from the impaired condition."

In each of these cases except the last, the physician in charge was either misinformed or failed to recommend a procedure that could have avoided a tragic outcome. In the last case, the testing laboratories were at fault. But these failures occurred at a time when the importance of genetic testing and genetic counseling was just beginning to be appreciated. It was also a time when the range of genetic diseases that could be detected in carriers was still quite limited. Today, the explosion in DNA technology and disease gene identification permits diagnosis of a wide range of genetic diseases both in newborns and parents contemplating having children. As discussed in Chapter 3, different ethnic groups recommend, and sometimes mandate, testing for genetic diseases that are prevalent among them.

This chapter begins with a discussion of the role of genetic counseling in preventing genetic disease. Had counseling been generally available at the time of these "wrongful life cases" as it is today, they might never have gone to litigation. The chapter then turns to the genesis of in vitro fertilization and how it made possible the selection of embryos lacking disease-producing mutations, thereby preventing the birth of children with genetic defects.

The role of genetic counseling

The Heredity Clinic at the University of Michigan, established in 1940 by Lee R. Dice, a geneticist and mammalogist at the university, was the first center to offer any sort of genetic counseling in the United States.[4] In 1948, James V. Neel, one of the founders of modern human genetics, became the physician in charge of the clinic. He was soon receiving letters from across the country requesting advice, but the clinic could only undertake the study of families who could actually make the trip to Ann Arbor. It shied away from giving advice by mail as this was akin to making a medical diagnosis by mail and clearly subject to error. The clinic was subsequently subsumed in the Department of Human Genetics, with the new department being chaired by Neel.

The Dight Institute at the University of Minnesota was the second counseling establishment to open.[5] The money to found the institute had been in the bequest of a physician named Charles Fremont Dight, a rabidly enthusiastic eugenicist and fan of involuntary sterilization of the "feebleminded." Between 1921 and 1935, the Minneapolis newspapers published some 300 letters written by Dight on various aspects of eugenics.

Dight was a rather odd chap. He lived for a time in an ingeniously designed tree house suitable for a single person. Most residents of Minneapolis probably remembered him more for his tree house than for his strident views on eugenics as it was the subject of innumerable newspaper articles and appears not to have been an entirely satisfactory way of living. Outside his tree house, Dight hung a sign that read, "Truth shall triumph. Justice shall be law." Dight managed to leave a sizable legacy to establish the Dight Institute by being very frugal, investing smartly, and failing to pay his income taxes. Dight wanted his eponymous institute "to promote biological race betterment—betterment in human brain structure and mental endowment and therefore behavior."

Dight died in 1937 and the Dight Institute for Human Genetics was founded on July 1, 1941.[6] In his will, Dight had stipulated that the institute that would bear his name and should provide courses and public lectures on human genetics, initiate research studies, and provide consultation on questions related to human genetics. Dispensation of eugenic advice and consultation was also encouraged. Its first director, Clarence P. Oliver, adhered to the advice given in Dight's will, but in 1946, he left to take a position at the University of Texas in Austin. Sheldon Reed succeeded him. It is to Reed we owe the term "genetic counseling" for Reed felt that expressions like "genetic hygiene" connoted things like toothpaste and deodorant, thus besmirching the serious nature of the endeavor. He also found that families he met with had little interest in eugenics and the term *genetic counseling* occurred to him "as an appropriate description of the process which I thought of as a kind of genetic social work" without eugenic connotations.

The Dight Institute Advisory Committee was not overly enamored of Reed's literary invention, but it has stood the test of time. Although Reed was not trained as a psychologist, he came to realize the importance of emotions like shame, guilt, blame, and resentment in the counseling process. Reed was instrumental in endowing genetic counseling with the ethos of respect and caring that characterizes the relationship between counselors and those they counsel today. In his classic text *Counseling in Medical Genetics* (1955), Reed wrote that "the second requirement" for a counselor "or perhaps the first—is that he have deep respect for the sensitivities, attitudes and reactions of the client."

According to the 1951 Dight Institute Bulletin, there were ten genetic-counseling centers in the United States at that time.[7] However, there were no educational programs for training genetic counselors. Today, the term describes somebody with a masters-level degree who has in-depth exposure to human genetics and counseling skills. The first program to meet this requirement was initiated at Sarah Lawrence College. It graduated its initial group of "genetic associates" in 1971. Melissa Richter, a biologist at the college who was also attracted to psychology, founded the program in 1969. She realized that counseling might provide an attractive career choice for women. Existing nursing and social work programs did not include genetic associate training within their domains, so Dr. Richter and her colleagues were able to convince medical professionals that counselors could serve a useful purpose, especially because human genetics was evolving rapidly. They proceeded to construct a well-designed program in genetics and counseling following consultation with geneticists.

In 1972, Dr. Richter went on a sabbatical to Radcliffe College, but she would never return to the program she had initiated for she was soon to die of breast cancer. Two codirectors succeeded her and one of them, Joan Marks, would lead the program over the next 26 years. The Sarah Lawrence program was directed at students who already had their bachelor's degree. They were taught the scientific principles of human and medical genetics as well as the fundamentals of counseling, including its important psychological dimensions. The first students to enroll in the program were older women who had raised their families and wanted to enter a new career.

Marks succeeded in attracting faculty members with expertise in areas such as human genetics, cytogenetics, biochemistry, and embryology to teach courses to her students. They were then introduced to families and patients with different genetic diseases at the several genetics clinics that existed in the greater New York area. Funding was always a problem because Sarah Lawrence College was not known for its scientific or medical orientation. Nevertheless, the program flourished and by 2004, when Marks received the American Society of Human Genetics Award for Excellence in Human Genetics Education, there were approximately 600 graduates of the program. They accounted for about one-third of the genetic counselors in the United States at the time. Practically all programs that have been initiated since were built along the lines of the Sarah Lawrence model and in 2004, one-half of these were directed by Sarah Lawrence graduates. Today, there are 29 accredited graduate programs in the United States and 3 in Canada.

It was only in 1977, when there were already several centers in the United States offering special training programs in genetic counseling, that the American Society of Human Genetics set up the American Board of Medical Genetics.[8] The Board was charged with accrediting training programs in genetic counseling. It was also tasked with creating a mechanism whereby professional credentials in medical genetics could be awarded to MDs and PhDs entering the field and also to genetic counselors with a master's certificate in genetic counseling.

In 1992, the American Board of Medical Genetics sought to become a member of the American Medical Association's Board of Medical Specialties. This meant that the American Board of Medical Genetics would have to exclude nondoctoral level candidates such as genetic counselors from the certification process. This resulted in the formation of the American Board of Genetic Counseling.[9] This Board currently has the responsibility for accrediting training programs for genetic counselors and also for certifying nascent genetic counselors.

Certification of a new counselor by the Board requires that the person has graduated from a Board-accredited training program; that he or she has acted as principal counselor for 50 supervised clinical cases during training; and that the counselor-to-be has passed two examinations. One of these stresses the general principles of medical genetics while the second focuses on counseling. The newly minted counselor will probably choose to join the National Society of Genetic Counselors that was formed in 1978 and publishes the *Journal of Genetic Counseling*.

Genetic counseling relates to all stages of the life cycle beginning with prospective parents who are concerned for some reason that their baby may be born with a genetic defect.[10] This could be because the prospective father or mother knows that he or she has a record of an inherited genetic disease in the family, cystic fibrosis for example; thinks that some congenital anomaly is present in the family, but does not know what it is; or the prospective parents are older and perhaps concerned about the likelihood that they may give birth to a baby with Down syndrome.

Counseling may also involve prenatal diagnosis. For example, heart defects are a leading cause of death in infants during the first year of life. If stillborn babies are included, around 30 per 1,000 babies have congenital heart defects, most of which are sporadic in origin.[11] If only live births are counted, the frequency of these sorts of defects are still quite high at 8 per 1,000. The fetal heart begins to form by day 18 and starts to beat by day 22. Performance of fetal echocardiography, an ultrasound technique, during the first trimester is used to detect irregular behavior of the heart. If it appears the fetus does have congenital heart disease, the counselor helps the family to understand the condition and to prepare for the birth of a baby with a diseased heart.

Down syndrome presents a different problem.[12] The chances of giving birth to a baby with Down syndrome rise dramatically with maternal age (refer to Table 2–3). Thus, about 30% of all Down babies are born to mothers 35 years of age or older. For expecting mothers younger than 35, the counselor will probably recommend maternal serum screening. This is a noninvasive procedure that usually involves testing for one protein (alpha-fetoprotein) and two hormones (unconjugated estriol and human chorionic gonadotropin). With Down syndrome, alpha-fetoprotein and unconjugated estriol are about 25% lower than in a normal pregnancy, whereas gonadotropin is about twice the normal amount. This test is usually performed in the second trimester of pregnancy between 15 and 18 weeks of gestation. The "triple test" will detect about 60% of Down pregnancies with a false positive rate of 9%. During the first or second trimester, ultrasound screening will detect a condition called nuchal translucency that results from the accumulation of fluid in the necks of Down syndrome fetuses. Combining ultrasound screening with the triple test probably increases the detection rate for Down syndrome fetuses to between 74% and 80%.

Should the triple test prove positive, counselors reassure the expecting couple that the test is not infallible and that a more invasive test such as chorionic villus sampling or amniocentesis may be in order. During the second trimester, fetal tissue can be obtained either by chorionic villus sampling (11–14 weeks), early amniocentesis (12–15 weeks), or second-trimester amniocentesis (15–20 weeks) for direct confirmation of Down syndrome by chromosome analysis. Chorionic villus sampling or amniocentesis is recommended for women 35 or older on their due date, although the counselor may suggest taking the triple test before reaching a decision about the other two options.

If it turns out that the couple is going to have a Down syndrome baby, the counselor provides them with up-to-date information about the condition and assists them in reaching a decision. They may choose to continue the pregnancy with the option of either raising the Down baby or putting it up for adoption or the couple may decide to terminate the pregnancy. As we have seen, the vast majority of couples choose to terminate the pregnancy (refer to Chapter 2, "How genetic diseases arise").

The decision to end a pregnancy means the couple has to find a doctor willing to perform abortions. But abortion is a very sensitive subject in the United States and abortion doctors risk their own lives in certain parts of this country, especially if they perform late-term or "partial birth" abortions. One such physician was Dr. George Tiller whose clinic was in Wichita, Kansas. Tiller became the preeminent abortion practitioner in America, advertising widely and drawing in women from all over the country.[13] He also drew frequent referrals from genetic counselors of expecting parents desiring to terminate Down syndrome pregnancies.

On Sunday, May 31, 2009, Reformation Lutheran Church was celebrating the Festival of the Pentecost with a special musical prelude. Most members, including Dr. Tiller's wife, were seated and the service was about to begin, but Dr. Tiller, an usher that day, was greeting stragglers in the foyer. Pastor Lowell Michelson was beating a goblet shaped drum known as a darbuka, most commonly found in the Middle East and parts of Africa, while singing an African song called "Celebrate the Journey!" All of a sudden, the pastor heard a sharp crack and was summoned to the foyer by another usher. George Tiller had been shot by an antiabortionist named Scott Roeder and died in the foyer.

Tiller's shooting was the climax of a long campaign of criminal harassment of Tiller by antiabortionists. For over 30 years, the antiabortion movement had been trying to drive him out of business thinking it would greatly cripple the "abortion industry." Women entering his clinic would often encounter protesters. They would see a "Truth Truck" whose side panels were adorned with large color photographs of dismembered fetuses. Children chanted "Tiller, Tiller, the baby killer," an expression picked up by Bill O'Reilly, the widely viewed Fox News host, who regularly attacked Dr. Tiller beginning in 2005.

On June 10, 2009, shortly after Tiller's death, Amy Goodman of Democracy Now interviewed a genetic counselor who wanted to remain anonymous. The counselor confirmed that the National Society of Genetic Counselors probably made "the vast majority of referrals" to Tiller's clinic. The counselor was distressed that the society had failed to make a public statement concerning Tiller's death. She inquired as to why this was the case and was informed "that there were some genetic counselors who specifically requested that they not make a public statement." She speculated that fear was probably the reason. But on June 14, the National Society of Genetic Counselors did issue a short, official statement condemning the murder. On July 1, 2010, Tiller's murderer, Scott Roeder, was sentenced to life in prison with no possibility of parole. Roeder had argued that he had chosen to obey "God's law" to save babies. As he said during his trial,

"I did kill him. It was not a murder. If you were to obey the higher power of God himself, you would acquit me."

Genetic counseling not only involves working with couples who are either thinking about having children or are already prospective parents, but also the diagnosis of genetic disorders in newborns, children, adult-onset disorders, and diseases such as Alzheimer's that usually affect the elderly.[14] A key process in genetic counseling is the establishment of a pedigree based on medical family history. The pedigree not only establishes the pattern of inheritance of a disease or condition, but it also helps to determine what genetic testing strategies to take. Construction of the pedigree is also an important tool for building a working relationship with the person or persons seeking advice. Depending on the nature of the condition or disease, the person being counseled may exhibit a range of emotions, including anger, anxiety, disbelief, denial, helplessness, and many other responses that the counselor will need to be able to cope with.

Suppose, for example, that a woman whose mother died from breast cancer consults a genetic counselor about her risk of contracting the disease. Although the vast majority of cases are sporadic, around 5%–10% are hereditary with the frequency of hereditary breast and ovarian cancer increasing in certain populations (refer to Chapter 3). The counselor will want to establish whether any of the woman's other female relations have had breast or ovarian cancer. If so, the counselor will probably suggest that the woman be checked for mutations in the *BRCA1* or *BRCA2* genes that can result in breast and ovarian cancer. Women with a *BRCA1* mutation have a 65%–85% lifetime risk of breast cancer and a 39%–60% lifetime risk of ovarian cancer. For a *BRCA2* mutation, the comparable risks are 45%–85% for breast cancer and 11%–25% for ovarian cancer.

Women at risk are advised to perform a monthly breast examination beginning at age 18 if a mutation has been detected that early in their lives. Beginning at age 25, monthly clinical breast examination and an annual mammogram are recommended. Transvaginal ultrasound and a specific antigen can be used to detect the appearance of ovarian cancer, although the efficacy of these methods is uncertain. Mastectomy and ovarian removal significantly reduce the risk of breast/ovarian cancer, but do not eliminate it. If a woman diagnosed with a *BRCA1* mutation wants to have children, she will have to risk contracting ovarian cancer until her children are born even if she has undergone mastectomy to reduce her chances of getting breast cancer. These are all topics she will want to discuss with her genetic counselor. The counselor can give her precise information on the risks she faces of contracting cancer and help her to reach a decision.

Now consider a different kind of problem. Suppose you have a couple that very much wants to have children, but they both have family members who suffer from cystic fibrosis. The counselor will probably suggest that they should be tested to see if they are carriers. If they are, there is a one in four chance that they will have a baby with the disease. To avoid this outcome, the counselor may suggest that the couple consider in vitro fertilization (IVF) followed by preimplantation genetic diagnosis (PGD). This will permit the couple to select for implantation only those embryos that lack the cystic fibrosis mutation or are carriers of the mutant gene. In this way, they can avoid giving birth to a child suffering from cystic fibrosis.

IVF presents a problem for certain religious groups. In a plenary address to the Congregation for the Doctrine of Faith on January 31, 2008, Pope Benedict XVI condemned the use of IVF saying that it had given rise to "new problems" such as "freezing of human embryos, embryonal reduction, pre-implantation diagnosis, stem cell research and attempts at human cloning." All these, said the pope, "clearly show how, with artificial insemination outside the body, the barrier to human dignity had been broken."[15]

Judie Brown, President of American Life League, the largest grassroots Catholic pro-life organization in the United States, was "elated" by the pope's condemnation of IVF, but was "saddened by the realization that the American Catholic bishops refuse to even take up an explanation of what the Church teaches let alone condemn the evil practice of in vitro fertilization." Consequently, most North American Catholics do not realize that IVF is immoral according to the teachings of the church.

In vitro fertilization (IVF)

On July 10, 2006, the *Daily Mail* reported that a young woman named Louise Brown was pregnant.[16] Ms. Brown was 27 and lived in Bristol, England, with her husband of two years Wesley Mullinder, a security officer. Ms. Brown and Mr. Mullinder made an unremarkable, middle-class English couple—except for one thing. Louise Brown was the world's first "test tube baby." Louise Brown's son Cameron was born on December 20, but, while the overjoyed mother said, "He's tiny, just under six pounds, but he's perfect," it was a bittersweet occasion as Louise Brown's father had died just two weeks earlier. So Louise Brown, conceived in a "test tube," had become pregnant the normal way and given birth to a little boy.

The story of in vitro fertilization really begins with Walter Heape, a physician and professor, at the University of Cambridge in England.[17] Heape was interested in the reproduction of different animal species. On April 27, 1890, he succeeded in transferring embryos from oviducts of an Angora rabbit female to the uterus of a recently mated Belgian hare. She produced four Belgians and two Angoras. So Heape had proved that it was possible to transfer preimplantation embryos from one pregnant female rabbit to another and have the donor embryos implant successfully in the uterus of the recipient.

Gregory Pincus, a Harvard endocrinologist remembered by most as one of the developers of "the pill," made the next major advance in the field when he and a colleague reported in 1939 that rabbit and human oocytes released from the follicles that contained them would undergo meiosis to yield eggs in vitro within 12 hours.[18] John Rock, one of America's best-known fertility specialists, followed Pincus. Rock, a committed Roman Catholic and a professor at Harvard Medical School, was highly skilled at tubal surgery. Yet tubal surgery only worked 7% of the time and was of no use to women who lacked Fallopian tubes or whose tubes were irreversibly damaged.

Rock came to believe that in vitro fertilization was the only hope in such cases. To accomplish this, he would have to extract an oocyte from a woman who was otherwise healthy, cause it to undergo maturation, fertilize it in vitro, and reimplant the resulting embryo in her womb. Rock retrieved more than 800 primary oocytes from women who had undergone hysterectomies. The state of maturation of the oocytes was crudely estimated and they were quite heterogeneous. Hoping that at least some of the oocytes had yielded eggs, Rock attempted to fertilize 138 of them in vitro. He thought he had observed embryonic cleavage in three cases suggesting that he had successfully fertilized these eggs. However, it is likely that Rock had not achieved fertilization and cell division at all because fragmentation of oocytes is fairly common in this work. He, of course, did not realize this. In 1944, he announced that he and his research assistant had managed to fertilize three human eggs in vitro and published his findings several years later.

The popular press learned of Rock's results and raised the hopes of infertile women that a solution to their problem was at hand. A 1947 article in *Parents* magazine reported optimistically, "nowadays specialists are effecting gratifying 'cures' of the apparently infertile." It was a typical case of the press getting ahead of the technology. Rock was besieged with requests from infertile women wishing that he could come to their rescue, not realizing that Rock was not even close to achieving IVF let alone embryo transfer. Some also offered their eggs and ovaries for additional experimentation.

In 1958, a young Edinburgh-educated embryologist, Robert Edwards, joined the National Institute of Medical Research. He was intrigued by Pincus's experiments and attempted to repeat them. Lo and behold, he found that oocytes liberated from mice, rats, and hamsters did mature to yield eggs within 12 hours, but those from humans, cows, sheep, rhesus monkeys, and baboons did not. Edwards spent two frustrating years varying conditions in order to get the latter oocytes to mature within 12 hours without success.

Molly Rose, a gynecologist who had delivered two of Edwards' daughters, provided Edwards human ovarian material from which he could extract oocytes, but she could only do so occasionally. Edwards was still unable to replicate Pincus's results and began to suspect that his timing was wrong for human oocytes. Perhaps they took longer. He waited for 18 hours for the oocytes to mature without success, but at 25 hours he had his eureka moment when the oocytes matured.

The next step was to achieve in vitro fertilization of human eggs, but for this Edwards needed a partner who could aspirate oocytes from their follicles in the ovary. Edwards knew that Patrick Steptoe, an obstetrician in the old textile-manufacturing town of Oldham near Manchester, had made use of a new fiber-optic device called a laparoscope to perform minimally invasive abdominal surgery. In 1968, Edwards phoned Steptoe and suggested that they collaborate on in vitro fertilization because he knew Steptoe could easily aspirate oocytes from their follicles. Edwards and Steptoe mulled over the safety and ethics of what they were about to do and agreed to participate as equals in their undertaking. They would stop if it appeared that their patients or children might be endangered, but not "for vague religious or political reasons." Edwards and Steptoe were partners for 20 years until Steptoe died in 1988.

Blocked Fallopian tubes were a major cause of female infertility and Steptoe realized that if he could aspirate eggs directly from infertile women, they could be fertilized in vitro and transferred into the uterus, thus avoiding the blocked Fallopian tubes. For about a decade, Steptoe and Edwards performed one in vitro experiment after another. Often, they used gonadotropin hormones to stimulate follicle formation and induce ovulation. They achieved some very short-lived pregnancies and one clinical pregnancy. But the latter was in the wrong place (ectopic) and had to be removed at 11 weeks. Ironically, the experiments carried out by Steptoe and Edwards were supported in part by money earned by Steptoe for performing legal abortions.

So far, they had waited four or more days before implantation by which time the embryo had undergone about 100 cell divisions. But after repeated failures, Steptoe and Edwards decided on a radical departure. They would introduce the embryo after just two and half days when it was at the eight-cell stage. Lesley Brown had blocked Fallopian tubes. She was only the second natural cycle patient that Steptoe and Edwards had worked with. All the others had been treated with gonadotropin to encourage follicle formation. Steptoe aspirated her single oocyte and quickly inseminated the resulting egg. The embryo was introduced into Lesley Brown's uterus at the eight-cell stage and she became pregnant. Nine months later, Dr. Steptoe delivered Louise Brown by Caesarean section on July 26, 1978. Her delighted father recalled, "It was like a dream. I couldn't believe it."[19]

Others were of a different opinion. University of Chicago biologist/ethicist Leon Kass was appalled and invoked a doomsday scenario worrying that "this blind assertion of will against our bodily nature—in contradiction of the meaning of human generation it seeks to control—can only lead to self-degradation and dehumanization." Paul Ramsay, a well-known Methodist ethicist, also shook his head and admonished the likes of Steptoe and Edwards. "Men ought not to play God before they learn to be men, and after they have learned to be men they will not play God." Predictably, the Roman Catholic Church was dismayed. "From a moral point of view procreation is deprived of its proper perfection when it is not desired as the fruit of the conjugal act, that is to say of the specific act of the spouses' union." Similar comments would accompany the embryonic stem cell debate at the beginning of the twenty-first century not to mention very recently with the successful implantation of a synthetic, but minimal bacterial genome.

The fact that IVF was likely to become popular with couples where one member of the pair was effectively infertile spurred governments into action.[20] In the United States, the 1973 *Roe v. Wade* decision was still quite recent as well as being extremely unpopular in the large part of the population that is opposed to abortion and fetal research. During the Carter administration, Health, Education, and Welfare Secretary Joseph Califano tasked a federal commission to make recommendations on reproductive services such as IVF. Many expected the commission to extend a moratorium on federal funding for this research that had been put in place late in the Nixon administration, but the commission instead reported positively. Despite the commission's recommendation, Califano and the National Institutes of Health were reluctant to provide funding for IVF research because of the extreme sensitivity of the topic.

In the private sector, however, there was great interest in IVF among infertile couples despite the expense, roughly $5,000 per cycle, and low probability of success, around 10% to 15% in 1987. However, success rates improved rapidly and private clinics multiplied. By 2006, there were 430 clinics in the United States and in that year, IVF was responsible for 43,412 births or around 1% of all babies born in this country. These clinics often provide donor sperm or eggs that can be used for IVF. They can even separate X chromosome- and Y chromosome-bearing sperm for sex selection.

The Centers for Disease Control monitors IVF and other assisted reproductive technologies (ART) under the 1992 Fertility Success Rate Act.[21] The act obliges the agency to provide annual reports on the topic, but does not allow the agency to exercise any sort of federal control over the practices of fertility clinics. The detailed annual reports contain a wealth of information. For example, the statistics show the overall success rate of IVF improved from 37% to 46% for women under 35 from 1998 to 2007, but drops markedly as maternal age increases (see Table 8–1). However, successful IVF frequently requires several attempts, each of which necessitates the administration of follicle-stimulating hormones. This causes the ovary to develop multiple follicles, each containing an egg. Normally, a woman produces a single egg per menstrual cycle, but many eggs are required to increase the probability of success.

Table 8–1 Percent success of IVF as a function of maternal age

	% success of IVF*	
Maternal age	**1998**	**2007**
Under 35	37	46
35–37	32	37
38–40	24	27
41–42	14	16

*These figures are for fresh, nondonor eggs.

Adapted from 2007 ART Report Section 5-ART Trends 1998–2007 (see Reference 21).

The British government has taken a different approach. In 1990, Parliament authorized the formation of the Human Fertilisation and Authority (HFEA), amending and updating the act in 2008.[22] Among other things, HFEA licenses and monitors fertility clinics; watches over establishments carrying out embryo research; maintains a register of licenses held by clinics, research laboratories, and storage centers; and regulates the storage of eggs, sperm, and embryos. This agency also deals with highly sensitive topics like embryonic stem cell research and therapeutic (as opposed to reproductive) cloning. However, in the United Kingdom, unlike in the United States, sperm selection for "family balancing" is banned.

A big problem concerns what to do with leftover stored embryos. Typically, multiple embryos are created for IVF and the best of these are used. This means that there are a lot of frozen embryos that will never be implanted. Over half of the embryos created in the UK between 1991 and 2005 (1.2 million) were never used.[23] Unused embryos in the UK may be discarded, frozen, donated for research, or donated to other infertile couples. Frozen embryos retained for future use must be destroyed after ten years.

In contrast, the United States lacks a federal authority regulating IVF and related technology. Instead, the Society for Assisted Reproductive Technology, which represents over 85% of the fertility clinics in the United States, works closely with the Centers for Disease Control to provide data on the outcomes of the procedures these clinics conduct.[24] However, this does not apply to precise statistics on stored embryos. A RAND Corporation study published in 2003 attempted an estimate of the number of stored embryos in the United States by surveying all 430 assisted reproduction clinics that existed in this country at the time.[25] Of these clinics, 340 reported having 396,526 embryos in storage as of April 11, 2002.

Similarly, there are no precise guidelines about how long stored embryos can be kept. A survey of embryo disposal practices elicited responses from 64% of 341 clinics contacted.[26] Virtually all were willing to create spare embryos and 175 of these practiced disposal whereas 33 did not. Of the clinics that did practice disposal, 78% required permission of both members of a couple. All clinics offered continued cryopreservation, although the majority (96%) charged a fee. Most clinics also offered to donate embryos for use by other clinics (76%) and for research (60%). Once a couple has successfully produced one or more children following IVF, the majority (72%) show little interest in the frozen embryos left behind.[27] But sooner or later, most of them have to confront the embryo storage problem and decide what to do because they receive a storage bill or a reminder from the fertility clinic.

The overall impression is that the British system of federal regulation is far more efficient than the scattershot approach taken in the United States. HFEA imposes standardized policies on IVF practices and clinics that are lacking in the United States. This applies not only to IVF, but also to embryonic stem cell research. HFEA defines the time at which human life commences as 14 days after fertilization when primitive streak formation begins. This is the point at which structural differentiation of the embryo commences. Embryonic stem cells are obtained from 3- to 5-day-old embryos at the blastocyst stage. In 2001, the Human Fertilisation and Embryology (Research Purposes) Regulations were enacted. These regulations effectively increased the scope of HFEA by extending the reasons for which an embryo could be created to include therapeutic cloning. There is no similar law in the United States. Reproductive cloning is banned in the UK as it is in the United States.

A good part of the reason that embryonic stem cell research fails to raise much concern in the UK is that the country is far more secular than the United States. On July 17, 2001, President George W. Bush, a practicing Methodist, restricted federal funding for stem cell research to 64 stem cell lines that existed at the time and specifically banned funding for research that required destruction of existing embryos to create stem cell lines or proposed the creation of new embryos for this purpose. In August 2002, the list was expanded to 78 lines that fitted the president's criteria. Although this might seem like a lot of stem cell lines, many were of doubtful quality or availability. The result was that privately funded initiatives such as the Harvard Stem Cell Institute came into existence or statewide initiatives such as California's, approved by voters in proposition 71, were enacted to provide state funding for embryonic stem cell research. On March 9, 2009, President Barack Obama signed an executive order lifting the ban on the use of federal dollars for embryonic stem cell research imposed by President Bush.

On October 4, 2010, it was announced that Robert G. Edwards, the man who started it all, had won the Nobel Prize in Physiology or Medicine. That same day, the International Federation of Catholic Medical Associations voiced its disagreement with the award.[28] The International Federation deplored the fact that IVF had led to creation of "millions of embryos" that would be discarded following their creation "as experimental animals destined for destruction." The statement went on to say that IVF "has led to a culture where [embryos] are regarded as commodities rather than the precious individuals which they are." And that is the central dilemma for many.

Preimplantation genetic diagnosis (PGD)

In 1967, Edwards made the first contribution to PGD when he showed that he could identify male and female rabbit embryos only five days after conception by using fluorescence microscopy to identify sex chromatin, the inactive X chromosome present in female cells, but not male cells (refer to Chapter 2).[29] But it required the arrival of another technique, the polymerase chain reaction (PCR) for PGD to really get off the ground. PCR permits the identification and amplification of specific DNA sequences starting with very small amounts of DNA (see Glossary).

In 1989, Alan H. Handyside, Robert Winston, and their colleagues at the Institute of Obstetrics and Gynaecology at Hammersmith Hospital in London reported they could extract single cells from embryos at the six- to ten-cell stage following which they used PCR to amplify a repeated sequence specific to the Y chromosome.[30] This allowed them to determine which of the embryos were male. The rest that lacked the sequence were assumed to be female. Handyside, Winston et al. pointed out that their approach "may be valuable for couples at risk of transmitting X-linked disease."

By the following year, 1990, they had demonstrated that this was indeed the case.

Five couples who were at risk of transmitting recessive X chromosome linked genes including X-linked mental retardation, adrenoleukodystrophy, Lesch-Nyhan syndrome, and Duchenne muscular dystrophy were informed that the sex of their embryos could be determined following IVF. The idea would be to discard the male embryos because they might be diseased and implant the female embryos. They might be disease carriers, but would themselves be disease free. At the eight-cell stage, two cells from each embryo were removed for duplicate PCR amplification using probes that would amplify a short repeat sequence present in the Y chromosome. Following transfer of the female embryos, pregnancies were confirmed in two and possibly three of the five women who had undergone treatment.

In 1992, Handyside and his colleagues extended their PGD technology to include the Δ508- deletion mutation that is responsible for some 70% of cystic fibrosis cases (Chapter 3).[31] Three women were involved this time. In the case of one poor woman, DNA analysis of one of the two embryos chosen for implantation failed, whereas the second proved to be homozygous for the deletion mutation so neither embryo was implanted. The fertilized eggs of the other two women formed carrier, noncarrier, and affected embryos. Both couples chose to have a noncarrier and a carrier embryo implanted. One woman became pregnant and gave birth to a girl free of the deletion mutation in both chromosomes.

The pioneering experiments of Handyside and Winston naturally attracted attention on the other side of the Atlantic. Yury Verlinsky was one of the first scientists in the United States to make use of PGD.[32] Verlinsky had been born in Ishim, Siberia, in the former Soviet Union. He received his PhD in embryology and cytogenetics from Kharkov University in the Ukraine, but he found it very difficult to obtain funding for his research in the Soviet Union. Verlinsky, his wife, and nine-year-old son immigrated to the United States in 1979. He soon obtained a research position at the Michael Reese Hospital in Chicago where he was put in charge of the cytogenetics laboratory because as he explained "of my experience and because a chromosome in any language is a chromosome and a microscope is a microscope."

In 1982, Verlinsky introduced chorionic villus sampling to the United States. While in the Soviet Union, he had worked on developing this method of fetal testing with a colleague, Anver Kuliev, with whom he had subsequently lost contact because interaction between émigré scientists and those remaining in the USSR was forbidden. Verlinsky later learned at a scientific meeting in Britain that Kuliev had had a brilliant career as a Soviet bureaucratic scientist and was now head of genetics at the World Health Organization. Verlinsky contacted Kuliev and together they began to contemplate the possibility of PGD, but it would be some years before the two scientists would be able to work together.

In the 1980s, Verlinsky also pioneered a technique that involved determining if, during egg formation, the egg had received a "good" or a "bad" chromosome with respect to a specific genetic disease. Human egg formation begins with the primary oocyte that is arrested in the first meiotic division (refer to Chapter 2). The primary oocyte completes the first meiotic division to form a secondary oocyte and a polar body. The secondary oocyte then enters the second meiotic division, but is arrested there until fertilized, whereupon it produces the fertilized egg and another polar body. Suppose you have a carrier of the Δ508 cystic fibrosis deletion. If the polar bodies contain the deletion, then the egg should not and ought to be suitable for IVF. By the early 1990s, Verlinsky was also using the methods pioneered by Handyside and Wilson that involved removing a cell from the embryo at the six- or eight-cell stage to see whether it contained the disease gene and if so whether it was a carrier.

By this time, Verlinsky had left the Michael Reese Hospital and started his own Reproductive Genetics Institute. Meanwhile, the Soviet Union collapsed and his old friend Kuliev joined him. The Reproductive Genetics Institute, housed in a two-story brick building in Chicago's Boystown, one of the largest gay and lesbian communities in the nation, was the first medical institute in the United States and one of the first in the world to offer chorionic villus sampling and PGD for genetic diseases. Verlinsky and his colleagues pioneered diagnostic tests for many genetic diseases so that today over 200 different genetic diseases and disorders can be detected using these methods.

Yury Verlinsky was diagnosed with colon cancer in 2007 and succumbed to the disease in July 2009. In Verlinsky's obituary in the *New York Times,* Dr. Andrew La Barbera, the scientific director for the American Society of Reproductive Medicine, was quoted as saying that Verlinsky "was a giant in the field because he transformed P.G.D. into a routine procedure that has enabled innumerable couples to conceive children free of genetic disease. He made it available to clinics around the world." Dr. Jamie Grifo, Program Director at the New York University Fertility Center, could only agree. "If it wasn't for Yury, who knows how far this field would have come?"

Today, Verlinsky's Reproductive Genetics Institute is a full-fledged fertility clinic. It is also one of the few that offers PGD for a great array of different genetic diseases.[33] It is the only one that uses Verlinsky's polar body method in addition to embryo screening.

Genesis Genetics Institute of Detroit, Michigan, also offers a wide variety of genetic tests.[34] It was founded by Mark Hughes, another of the pioneers in PGD. Today, Genesis Genetics is the PGD laboratory of choice for many fertility clinics worldwide.

There is another use for IVF combined with PGD. Suppose a couple has a child with a nasty genetic disease like Fanconi's anemia. It is a disease that can be alleviated with a bone marrow transplant from a healthy donor or by using umbilical cord blood from a newborn, but it is important that the donor is well matched immunologically to the recipient. In the absence of a suitable brother or sister, the parents of the afflicted child may choose to make use of IVF and PGD to select an embryo for implantation into the child's mother that is disease free. Both Verlinsky and Hughes have made use of these methods, as we shall see in the next chapter, to try to help families with Fanconi children to obtain suitable matches.

IVF can be expensive and prices vary (see Table 8–2). Furthermore, a single cycle of IVF is often not sufficient to produce a pregnancy. Thus, the Genetics & IVF Institute of Fairfax, Virginia, offers a single cycle of IVF for $8,900.[35] However, their Delivery Promise program, which offers four cycles of IVF plus a refund if unsuccessful, costs $19,500 for women under 36 and $24,500 for women 36 to 38 years old. PGD for a limited number of genetic diseases that includes cystic fibrosis and sickle cell anemia costs an additional $5,000. Furthermore, insurance coverage of IVF, except in states like Massachusetts, Connecticut, and Rhode Island, which have passed laws mandating that insurers help pay the cost of in vitro fertilization, is highly variable and might not be offered at all.[36]

Table 8–2 The cost of a single cycle of in vitro fertilization*

Clinic	Location	Cost of a single cycle
Genetics & IVF Institute	Virginia	$8,900
NYU Fertility Center	New York	10,565
Center for Reproductive Medicine & Advanced Reproductive Technologies	Minnesota	12,442
The Fertility Center	Michigan	5,542
North Houston Center for Reproductive Medicine	Texas	9,878

*Costs listed on the Web sites of each clinic as of November 2010. These clinics and a host of others listed alphabetically for each state can be found on IHR.com Infertility Resources for Consumers. http://www.ihr.com/infertility/provider/.

9

Treating genetic disease

Robert Guthrie was born in the Ozarks in 1916.[1] He received both an MD and a PhD from the University of Minnesota and eventually joined the Department of Pediatrics where he was appointed Professor of Pediatrics and Microbiology. Guthrie mainly credited his niece, Margaret Doll, for inspiring him to campaign for newborn genetic testing and, perhaps even more significantly, for developing a method for detecting phenylketoneuria in infants. His niece, who was profoundly mentally retarded, was diagnosed as having phenylketoneuria at the age of 15 months, too late to prevent brain damage.

Guthrie succeeded in developing a simple test that would make screening of newborns for phenylketoneuria rapid and accurate.[2] He took spores of a bacterium called *Bacillus subtilis* and mixed them in a medium containing thienylalanine, an analog of phenylalanine, that would prevent their growth. Once the agar-containing medium had hardened, he could add filters impregnated with newborn blood to the medium. If the blood contained phenylalanine, the effect of the analog was reversed and the cells grew. Overnight incubation was sufficient to obtain results. Guthrie's test was soon widely adopted, the first such test for a genetic disease in newborns.

After the development of the Guthrie test, a number of states and municipalities initiated testing programs for phenylketoneuria.[3] In New York City, 51 newborns with phenylketoneuria were identified between 1966 and 1974. The price of testing infants came to less than a dollar each, adding up to no more than a million dollars, whereas the expense of keeping these children with phenylketoneuria institutionalized would probably have been in the neighborhood of 13 million dollars.

The obvious cost effectiveness of the New York and other phenylketoneuria screening programs suggested that the scope of newborn screening should be increased to include other genetic diseases. By 2002, almost every baby born in an American hospital, birthing center, or at home with a midwife was contributing a few drops of blood to be screened for phenylketoneuria and two or more other serious hereditary disorders.[4] Although every state now had a newborn-screening program, the number of disorders screened varied between states. In 2002, Wisconsin boasted the most advanced newborn-screening program, testing for 21 inherited diseases, followed by Massachusetts, which tested for 11, but dropping to just 3 for babies born in West Virginia.

Dr. Kenneth A. Pass, who directed New York's Newborn Screening Program in Albany, reported that 1 baby in 350 was positive for one of the eight diseases tested for in that state. Dr. Pass pointed out in a seminar sponsored by the March of Dimes, whose mission is to improve the health of babies by preventing birth defects, premature birth, and infant mortality, that "Although the conditions, taken individually, are rare, collectively they are not so rare." Then Dr. Pass made a telling point saying that as the technology had improved "the economics of screening have increased since more information can be gleaned from the same five drops of blood at no additional cost."

In the spring of 2005, a federally appointed advisory group issued a report recommending that a uniform method of screening of newborns for 29 different genetic diseases be adopted across the United States.[5] For five or six of these conditions, it is unclear whether the treatments alleviate the problem or the frequency with which a baby may exhibit a false positive result. Furthermore, the tests unintentionally identify about 25 other conditions besides the 29 they were intended to target. How significant these other conditions are or whether they are associated with specific genetic diseases is or was at the time unknown. For these reasons, reactions to the report varied greatly, particularly with respect to how the information obtained should be disseminated and used.

"Giving parents the result, saying, 'Here's the mutation; we are not sure what the outcome will be,' is better than not telling," according to Sharon Terry, President and Chief Executive Officer of the

Genetic Alliance, a group that promotes optimum health care for people with genetic disorders. Ms. Terry felt it was paternalistic for doctors to assume that it was better for parents not to know.

Dr. R. Rodney Howell, a pediatrics professor at the Leonard M. Miller School of Medicine at the University of Miami, who chaired the committee that wrote the report as well as the federal advisory group, agreed.

"Do I feel it will be difficult for physicians and caretakers to deal with this?" Dr. Howell remarked. "The answer is yes. But I just don't think it is proper for us to have information about an anomaly without conveying it."

But Dr. Lainie Friedman Ross, a University of Chicago pediatrician and ethicist, worried: "We don't know if they are medical conditions. We don't know what to do with the information. Reporting test data for which there are no systems in place for follow-up testing and treatment is not rejecting paternalism, but it is patient abandonment."

The disparate nature of these responses is understandable. If a newborn has a genetic disease, it is important that parents be informed immediately of the nature of the disease, its consequences, and its treatment. But if it has 1 of 25 ill-defined conditions that might or might not be important, why worry the parents about it unless there is something that can be done to alleviate the condition?

Whatever the ethical reservations people might have, Dr. Howell noted that the states were expanding testing programs rapidly: "It's not really a question of, 'Should we expand newborn screening?' It's happening. It's going like a house on fire."

This is certainly the case. By March 2010, all 29 conditions were being screened for in all 50 states with the exception of 1 or 2 conditions in 1 or 2 states.[6] Furthermore, the report on newborn screening listed 25 other genetic conditions that were of secondary concern. Many of these are also being tested for in most states.

In December 2008, President George W. Bush's Council on Bioethics delivered the president a lengthy white paper on newborn screening.[7] In his cover letter, Chairman Edmund D. Pellegrino conveyed the council's recommendation "that the states mandate newborn screening only for diseases that meet traditional criteria, including the availability of an effective treatment." This would not preclude the participation of infants in pilot studies for "conditions that do not meet the traditional criteria," but this should only be with the "informed consent of the infant's parents." The council felt "that the potential benefits of mandatory, population-wide newborn screening for diseases for which there is no current treatment are outweighed by the potential harms."

In October 2009, Francis Collins, Director of the National Institutes of Health, said, at a meeting organized by the American Association for the Advancement of Science, that "whether you like it or not, a complete sequencing of newborns is not far away."[8] Dr. Collins may well be right and all sorts of risk-factor genes are likely to be identified in addition to the occasional disease that needs to be treated soon after birth. But what are parents, or for that matter the child after it grows up, going to with this information? It will simply worry them needlessly unless the meaning of these risk factors is clearly explained by a specialist in medical genetics such as a counselor or doctor with the appropriate training.

Newborn genetic testing also yields leftover samples that are sometimes employed for research without parental consent or knowledge.[9] These samples are not only used to improve screening methods, but also to do research on bigger problems such as the nature of genetic alterations that cause childhood cancer. Texas has been forced to dispose of samples from more than 5 million babies to settle a lawsuit brought by parents who objected to secret DNA warehousing. The judge dismissed a similar suit brought in Minnesota. Michigan incorporated 4 million leftover blood spots into a new "BioTrust for Health." Public health officials plan a campaign to educate parents on the virtues of allowing retention of the samples for research purposes, but also to explain to families how they can opt out. Federal government advisers are working on national recommendations to ensure that all newborns still get tested, but allowing parents to have more influence on what happens to the samples following testing.

Similar concerns have been raised in the United Kingdom where some facilities have kept samples for as long as 20 years despite the fact that the government advises that the samples be destroyed after 5 years.[10] Even though the DNA of each newborn is stored anonymously, the UK Newborn Screening Programme Centre, which oversees the use of these samples, points out that they could be linked to hospital admissions and a child could be identified in that way. Shami Chakrabarti, the director of Liberty, an organization devoted to protecting civil liberties and promoting human rights, said that as "someone who gave consent for my own baby to be tested, I'm horrified that anyone would breach my trust, keep my child's sample for years on end and use it for all sorts of extraneous purposes." She wrote the health secretary on behalf of Liberty asking for an immediate investigation.

There is no doubt that newborn screening serves a very important purpose, but it is equally true that parents should be informed and their permission requested if their baby's blood samples are to be kept for use in research. In this regard, the Michigan BioTrust for Health offers an appropriate model.[11] The Web site poses a series of questions and answers that explain why there are leftover blood spots, how they are stored, and how they are used in research. The Web site also provides instructions for parents who do not want their child's blood spots used in research. "If you or your child were born between July 1984 and 2010, there are two options. You may fill out a form to: (1) request that the sample be saved but not used for research, or (2) request that the sample be destroyed after newborn screening is completed."

Newborn screening is the first line of defense in combating genetic disease and, as we shall see in the next section, the vast majority of the genetic defects screened for in the panel of 29 primary disease targets are ones in which dietary modifications can make a huge difference (see Table 9–1). However, speed is of the essence.

Table 9–1 Treatments for newborn genetic diseases*

Disease	Frequency x 10^4	No. of Genes	Treatment
Congenital hypothyroidism (TCH)	2.5–3.3	7	Thyroxine
Congenital Adrenal hyperplasia (CAH)	1–0.6	1	Cortisol
Biotinidase (BIO)	0.17	1	Dietary
Galactosemia (GALT)	0.17–0.33	3	Dietary
Carnitineuptake deficiency (CUD)	0.1	1	Dietary
Long-chain L-3-Hydroxyacyl-CoA-Dehydrogenase (LCHAD)	0.16	1	Dietary in part
Medium-chain acyl-CoA dehydrogenase (MCAD)	0.6	1	Dietary
Trifunctional protein (TFP)	?	2	Dietary in part
Very long-chain Aclyl-CoA Dehydrogenase (VLCAD)	.08–.25	1	Dietary
Glutaric acidemia Type 1 (GA)	0.33	1	Dietary
3-hydroxy 3-methyl glutaric aciduria (HMG)	very rare	1	Dietary
Isovaleric acidemia (IVA)	0.04	1	Dietary
3-methylcrotonyl-CoA carboxylase (3-MCC)	0.28	2	Dietary
Methylmalonic Acidemia (Vitamin B12 disorders) (CBL A,B)	0.10–0.20	3	Dietary

Disease	Frequency x 10^4	No. of Genes	Treatment
Methylmalonic Acidemia (MUT)	0.1–0.2	3	Dietary
Beta ketothiolase (BKT)	0.01	1	Dietary
Propionic academia (PROP)	0.10	2	Dietary
Multiple carboxylase (MCD)	0.11	1	Dietary
Argininosuccinate Aciduria (ASA)	0.14	1	Dietary
Citrullinemia Type 1 (CIT1)	0.175	1	Dietary
Homocystinueria (HCY)	0.03–0.05	4	Vitamin B6
Maple syrup urine disease (MSUD)	0.05	4	Dietary
Phenylketoneuria (PKU)	0.67–1.0	1	Dietary
Tyrosenimia Type 1 (TYR-1)	0.25	1	Dietary

Omitted are tests for hearing, cystic fibrosis, sickle cell anemia and trait, and severe combined immunodeficiency. Genetic disease abbreviations are standard.

*Most of the information in this table is derived from Genetics Home Reference (http://ghr.nlm.nih.gov/) and references, particularly MedLine Plus contained therein. The table includes 24 of the 29 gene defects screened for in newborns.

The importance of diet

The 24 genetic diseases listed in Table 9–1 are among the 29 primary target diseases that are screened for in newborns. Like PKU, the vast majority are metabolic diseases that respond at least in part to dietary therapy.

One of these diseases, called maple syrup urine disease (MSUD), has a rather long and interesting history that involves Robert Guthrie once more.[12] In 1951, Dr. John Menkes, an intern at Boston Children's Hospital who was planning a career in pediatric neurology, encountered a newborn infant who was drowsy and who also refused formula. The baby had been diagnosed as having kernicterus, a kind of brain damage that results from jaundice.

Menkes discussed the baby's condition with its mother and learned she had earlier had two other children whose urine had an unusual odor and had died, whereas a daughter whose urine lacked the odor had lived. By the third day, the urine and perspiration of the infant that had been brought to Menkes' attention also had a strange odor.

"I am sure that I must have asked nearly everyone at the hospital but could get no better answer than that it smelled like maple syrup," Dr. Menkes recalled later on. "After all, we were in New England."

Why was it that the urine of infants with MSUD smelled like maple syrup? The answer to this question came in the late 1950s from Menkes, and from a collaborative venture between Joseph Dancis at New York University-Bellevue Medical Center and Roland Westall at University College Hospital Medical School in London.[13] They independently discovered that there were increased levels of three amino acids, leucine, isoleucine, and valine, in the baby's urine and plasma. This suggested that there was a block in the pathway leading to the degradation of these three amino acids. This caused the accumulation of breakdown products called keto acids.

Like phenylalanine, all three amino acids are essential. That is, humans cannot synthesize them. This meant that dietary control might work for MSUD as it did for PKU. Roland Westfall and his colleague C. E. Dent at University College announced progress on this front in 1961. They attempted dietary modification on a girl with MSUD who had been transferred to University College Hospital at the age of eight months. She was already severely mentally retarded and lay on her back with only occasional attempts to raise her head. Westfall and Dent were able to devise a diet that resulted in normal plasma levels of isoleucine, leucine, and valine and also caused the urine to lose the odor of maple syrup. Furthermore, the baby gained weight, but her clinical condition did not change. Other dietary improvements followed.

Meanwhile, Robert Guthrie and his colleague Edwin Naylor had come up with a bacterial screening test for MSUD similar to the one they had devised for PKU.[14] They began a field trial of the assay in July 1964 together with five laboratories that were already collaborating with them on PKU screening. These laboratories, all in the United States except for one in New Zealand, reported their results monthly to Naylor and Guthrie. They were also getting annual updates from laboratories around the world. By 1978, more than 9 1/2 million newborns had been screened for MSUD worldwide.

Today, we know that mutations in any one of four genes can lead to MSUD.[15] Each gene encodes a protein, forming part of an enzyme complex that is responsible for processing the keto acid breakdown products of leucine, isoleucine, and valine to the next step in the degradation process. The diagnostic process involves recognition of the characteristic clinical features of MSUD, but the Guthrie test is no longer used to assay for amino acid accumulation. Instead, the activity of the enzyme complex is measured directly or else molecular genetic testing of all four genes may be done. MSUD is managed by using high-calorie formulas free of leucine, isoleucine, and valine. Because all three amino acids are required for protein synthesis, they must be supplied, but in carefully monitored amounts.

The 29 primary genetic diseases targeted for newborn testing include several other amino acid disorders, organic acid disorders, fatty acid disorders, plus several other conditions (refer to Table 9–1). Treatment of these diseases includes dietary modification, but this is often not sufficient by itself. Consider tyrosinemia type I (TYR-1). As is true of PKU and MSUD, this disease involves a defect in the degradation of an amino acid, in this case tyrosine.[16] The genetic defect in this disease causes the terminal enzyme in the tyrosine degradation pathway to lose function. This means that fumarylacetoacetic acid

(FAA), the precursor with which this enzyme reacts, begins to accumulate. FAA and other compounds derived from it ultimately cause severe liver damage. If left untreated, this means the child will usually die before the age of ten.

As soon as TYR-1 is diagnosed, a compound called nitisinone is given. Nitisinone blocks the tyrosine degradation pathway at an earlier step. This prevents FAA accumulation and subsequent liver damage. However, nitisinone treatment causes tyrosine to accumulate in the blood. This means that the patient must be kept on a low tyrosine diet to prevent the accumulation of tyrosine crystals in the cornea.

Hormones and proteins

Congenital hypothyroidism (TCH) and congenital adrenal hyperplasia (CAH) are among the panel of genetic diseases tested for in newborns (refer to Table 9–1).[17] Both diseases result in hormone deficiencies that can be corrected with administration of the appropriate hormones. TCH arises when the thyroid gland fails to develop or function properly. Even though mutations in seven different genes have been found to cause TCH, the vast majority of cases are sporadic and, therefore, not predicted on a familial basis. Diagnosis of TCH involves measuring two thyroid hormones, the thyroid stimulating hormone and thyroxine. The test is best carried out on blood droplets collected when the baby is between two and six days old. Rapid diagnosis is important because, in the absence of these hormones, IQ begins to decline noticeably and mental retardation results.

Congenital renal hyperplasia usually arises as a consequence of mutations in the *CYP21A* gene.[18] This gene encodes an enzyme called 21-hydroxylase. The enzyme is found in the adrenal gland where it is involved in producing the hormones cortisol and aldosterone. Cortisol plays an important role in maintaining blood sugar levels, protecting against stress, and suppressing inflammation. Aldosterone regulates salt retention by the kidneys.

A deficiency of 21-hydroxylase leads to the accumulation of compounds that are normally converted to cortisol and aldosterone. These substances then build up in the adrenal glands and are converted into male-determining hormones or androgens. Consequently, females with this disease usually have external genitalia that look neither distinctively male nor female. They may also exhibit the growth of excess body hair, male pattern baldness, and unpredictable menstruation. About 75% of affected individuals have the most severe form of the disease that is known as the salt-wasting type. This causes the loss of large amounts of sodium in the urine. Infants with this form of the disease feed poorly, lose weight, and are subject to dehydration and vomiting. The disease is life threatening if not treated. Fortunately, by supplying a form of cortisol, aldosterone, or both, the symptoms are readily reversed. The hormone(s) must be supplied for life, but the prognosis is good.

Hemophilia is probably the most familiar genetic disease treatable by a protein. Mutations in the sex-linked *F8* and *F9* genes, respectively, cause hemophilia A and B. *F8* encodes clotting factor VIII, whereas *F9* specifies clotting factor IX (also refer to Chapter 1, "Hunting for disease genes," with reference to hemophilia in Queen Victoria's progeny).[19] By treating hemophiliacs at regular intervals with the appropriate clotting factor purified from blood plasma, bleeding can be controlled. Current guidelines recommend the infusion of concentrated factor VIII three times a week for hemophilia A and of factor IX twice a week for hemophilia B. The concentrates can be administered at home.

But there is (or was) a potential problem. To purify factors VIII and IX blood, concentrates are pooled from many donors. If the blood of any one of these donors is infected with a virus, that virus can be transmitted to the hemophilic recipient. The reason why hemophilia is probably familiar to many is because many hemophiliacs became HIV positive in the 1970s and 1980s following treatment because the concentrates used were contaminated with the AIDS virus.[20]

After the discovery of the AIDS virus in 1981, it became evident that homosexual men and intravenous drug users were its main victims, but in 1982 the Centers for Disease Control reported that three hemophiliacs had become ill from the AIDS virus. Because HIV was prevalent among intravenous drug users and homosexuals, some of whom were blood donors, in March 1983, the agency warned that the virus might be contaminating blood products, including the concentrates used for hemophilia prevention. However, there was no way of telling which donors were HIV positive or which plasma samples contained the virus because a test for its presence had yet to be developed.

Cutter laboratories, a division of Bayer, was a producer of concentrates for hemophiliacs and in January 1983, its manager for plasma procurement acknowledged that there was good reason to believe that AIDS was being transmitted to other people through the use of plasma products.[21] Then, Cutter learned that a rival was heating the concentrate to inactivate the virus. Cutter's sales were slipping, but, according to an internal memo, rather than admitting that they had not so far supplied heated concentrate, Cutter conceived a marketing plan that was designed to give the impression that they were continuing to improve their product without acknowledging that they soon expected to have heated concentrate available.

In June 1983, the company tried to minimize the danger hemophiliacs faced from using their unheated concentrate. In a letter to distributors in 21 countries, Cutter wrote, "AIDS has become the center of irrational response in many countries. This is of particular concern to us because of unsubstantiated speculation that this syndrome may be transmitted by certain blood products."

The consequences were particularly egregious in France, where nearly half of the country's 3,000 hemophiliacs contracted AIDS.[22] In a series of articles published in *L'Evenement du Jeudi* in the spring of 1990, physician-journalist Ann-Marie Casteret revealed that the reason so many French hemophiliacs had contracted AIDS could be traced to a decision made in 1985 by officials at the French National Center for Blood (CNTS).

In October 1984, Michel Garetta became director of the CNTS. The organization had been operating at a loss and was dependent on the Ministry of Health to make up the difference, so anything Garetta could do to make the organization more profitable would be helpful. Furthermore, an important goal of the CNTS was to make France self-sufficient in blood products. So Garetta stepped up the production of factor VIII without taking the time to develop a heating method, even though American laboratories like Travenol had began advising as early as 1983 that it was best to use heated concentrate for hemophiliacs. It was a terrible error.

Garetta's decision became even more inexcusable in view of an internal memorandum sent to the health ministry by Dr. Jean François Pinon in January 1985. Pinon was chief of hematology at the Cochin Hospital in Paris. An immunologist, Jacques Leibovitch, had convinced him it might be a good idea to screen the hospital's stocks of blood. On December 12, Leibovitch reported that 10 out of the 2,000 samples he had tested were positive for HIV. The implications for Garetta were devastating. Because CNTS produced its concentrates for factor VIII production from up to 5,000 pooled blood donations, all were almost certainly contaminated.

In February 1985, Luc Montagnier, discoverer of the AIDS virus, reported in the *Lancet* that heat-treating blood destroyed the virus and that heating the concentrate prevented infection. Meanwhile, in the same month, both Pasteur Diagnostics in France and Abbott Laboratories in the United States filed applications in France for AIDS virus tests that they had developed independently. The Abbott test was approved for use in the United States in March and by April Robert Netter, Director of the French National Health Laboratory, was urging approval of the Pasteur test because he did not see how approval of the Abbott test could be delayed much longer without an appeal by Abbott to the Council of State charging abuse of power.

In a French cabinet meeting held on May 9, 1985, chaired by François Gros, Health Adviser to Prime Minister Laurent Fabius and a former Director of the Pasteur Institute, the decision was made to continue to hold off on the testing of blood donors for AIDS using either the Abbott or Pasteur tests on the basis that it would cost too much. Furthermore, recommendations to heat-treat both the French and imported bloodstocks were ignored. Finally, in July 1985, both the Abbott and Pasteur tests were approved.

It was also on May 9, 1985, months after Pinon's warning, when Garetta finally contacted the Ministry of Social Affairs saying that purging the French blood supply of HIV contamination was essential. However, Garetta continued, a compromise was necessary between the interests of public health and economic realities. From that time, each delay of three months in testing donors and preparing heated factor VIII would result in the deaths of five to ten hemophiliacs unable to obtain factor VIII to correct bleeding episodes. Furthermore, the economic consequences of having to import heated factor VIII and destroy contaminated stocks coupled with the 20% production loss that occurred because of heating would cost millions of francs (in dollar terms, well over $5 million). Those costs would have to be covered by the Ministries of Health and Social Services.

To minimize all of these problems, Garetta recommended that the CNTS continue to provide unheated stocks to its clients until mid-July for what were mainly economic reasons. The Association of French Hemophiliacs went along with Garetta's recommendation, but urged that the ministry ban unheated products after October 1, 1985. Why would they have done so? Because Garetta had told the association that technical problems had held up the production of heated product by the CNTS while omitting to mention the fact that factor VIII stocks were contaminated.

Casteret's 1990 articles triggered a government investigation. Michel Lucas, the Inspector General of Social Affairs, spent three month locating and studying official documents on the contaminated blood scandal. He constructed a precise chronology of the events that led to the conclusion that the decision to continue using the suspect blood for six months without an AIDS test was monetary. In 1991, Garetta and three other health officials were put on trial before a panel of judges at the Palais de Justice in Paris. Garetta was sentenced to four years in prison with a substantial fine. He was released a year early for good behavior.

But the investigation did not end there. On February 10, 1999, former Prime Minister Laurent Fabius and two other ministers in the government he headed from 1984 to 1986 were put on trial in connection with the contaminated blood scandal that had occurred during their administration. On March 9, Fabius and his Social Affairs Minister, Georgina Dufoix, were acquitted of manslaughter. The former Health Minister, Edmond Herve, was convicted of two cases of negligence, but was not sentenced. The court ruled that Herve should have ordered the destruction of the untreated blood samples once it was known that they might contain the AIDS virus. He faced a possible five-year sentence, but the court found that the scandal had gone on for so many years that Herve had been denied the right of presumption of innocence. The French Transfusion Association, an advocacy group for transfusion victims, called the acquittals "disgraceful."

Meanwhile back in the United States, Cutter got approval to sell heated concentrate on February 29, 1984, the last of the four major companies that distributed blood products to do so.[23] But Cutter continued to make the older unheated product for five more months in order to profit from several large contracts where the price had already been fixed, and also because some customers wanted the unheated concentrate believing that heating might cause the concentrate to deteriorate.

In a meeting held on November 15, 1984, Cutter noted that it had an excess of unheated concentrate. Because Cutter had incurred considerable costs in obtaining, preparing, and storing the concentrate, the company wanted to recoup its expenses somehow. At the same meeting, company executives decided to see whether the additional product could be sold abroad. Cutter successfully sold more than 100,000 vials of the unheated concentrate worth over $4 million, particularly in Asia. Doctors in Taiwan, Singapore, Malaysia, and Indonesia were primarily dispensing the unheated concentrate.

At Hong Kong's Queen Mary Hospital, Dr. Chan Tai-kwong lamented that he and other doctors at the hospital were unable to obtain the heated concentrate from Cutter. Dr. Chan found that 40% of his patients were HIV positive. Dr. Patrick Yuen, who was associated with a different hospital, reported similar results. The story was repeated in Taiwan. A scandal like the one in France also erupted in Japan.

AIDS virus is not the only virus that may contaminate concentrates of factor VIII.[24] Since the 1970s, when plasma concentrates of coagulation factors first became widely available, hepatitis B and C viruses have occasionally been unwelcome contaminants. Today's concentrates prepared from plasma are much safer. In addition to heating, a solvent-detergent mixture is now used to prevent viral infection. The latter mixture inactivates viruses with a lipid envelope, but does not work for the hepatitis A virus that lacks such an envelope.

The early 1990s also saw the licensing of the first two recombinant clotting factors. These factors are generally produced in hamster cells in tissue culture. Currently, four preparations are available.[25] Bayer makes two of these, Helixate FS and Kogenate FS. Advate is a Baxter product and Wyeth produces Xyntha.

Treating hemophilia is expensive because factor VIII concentrates, whether obtained from pooled blood samples or produced using recombinant DNA, cost a lot. In one recent study of boys with severe hemophilia treated with Kogenate, the estimated cost per child per year was in the neighborhood of $300,000.[26] The National Hemophilia Association estimates the cost of treatment per person per annum at $60,000 to $150,000.[27] The association seemingly throws up its hands where insurance is concerned. "The state of insurance reimbursement today is constantly changing to meet this competitive environment. Patients and medical professionals alike are finding the system difficult to manage, complex and often confusing. While organizations and treatment professionals can educate people to better understand insurance, it ultimately falls on individuals to manage their own healthcare reimbursement."

Enzyme replacement therapy is another method for alleviating certain genetic diseases, particularly the lysosomal storage diseases.[28] These tend to be at higher frequency in Ashkenazi Jewish populations. The lysosome, a membrane-bound organelle, is tasked with degrading various cellular waste products or digesting substances brought in from outside the cell. Loss of lysosomal enzyme function promotes the accumulation of substrates for these enzymes. This, in turn, results in cellular death, and tissue damage or organ failure.

Gaucher disease is the one of the most common lysosomal storage diseases. It results from a deficiency in an enzyme called acid beta-glucosidase. As a consequence, glucocerebroside accumulates in the lysosomes of cells called macrophages. Macrophages are produced by differentiation of certain kinds of white blood cells called monocytes. They are capable of amoeboid movement and ingest cellular debris as well as pathogens. Digestion of a pathogen involves formation of a structure called a phagosome by the macrophage. Fusion of lysosomes with the phagosome creates a phagolysosome in which the pathogen is broken down by lysosomal enzymes. In Gaucher disease, these cells become engorged because they cannot eliminate glucocerebroside and assume a characteristic crumpled tissue paper appearance that has led to them being called Gaucher cells.

Although there are three types of Gaucher disease, Type 1 accounts for 80% to 90% of all cases. In the majority of individuals with this disease, its symptoms are not expressed until adulthood. Type 1 Gaucher disease is currently a target for enzyme replacement therapy. It turns out that, during the normal cellular processing of a lysosomal enzyme, a tag is attached to the enzyme molecule that is made up of a sugar molecule (mannose) with a phosphate residue (mannose-6-phosphate). A mannose-6-phosphate receptor molecule then sorts the enzyme molecule into an intracellular transport vesicle that is destined for a lysosome.

The whole cellular processing sequence that the enzyme molecule undergoes before it enters a lysosome cannot be mimicked by simply injecting a Gaucher patient with enzyme. Instead, enzyme therapy takes advantage of the fact that macrophages have mannose-6-phosphate receptors. A recombinant enzyme is used that has the mannose-6-phosphate tag. This results in its uptake by macrophages and its transfer to lysosomes within the macrophage. The enzyme can then catalyze the breakdown of the accumulated glucocerebroside. The FDA has approved this general enzyme replacement strategy for Gaucher diseases and for several other lysosomal disorders.

The main problem with treating Gaucher disease using enzyme replacement therapy is the expense.[29] Treatment was estimated to cost $145,000 to $290,000 per year in 2005. Estimates for drugs used to treat three other lysosomal storage diseases that were approved at that time ranged from $156,000 per year to $377,000 per year. The reason is that treatment for Gaucher disease and related lysosomal storage diseases is covered by the Orphan Drug Act of 1983.[30] Its specific purpose was to encourage pharmaceutical companies to develop drugs that alleviated rare diseases. One of the rewards for a company that agreed to do this was a seven-year exclusivity clause upon FDA approval. Once a company was approved to create a certain drug, it was unlikely that other pharmaceutical companies were going to develop competing drugs anyway in view of the small number of patients involved and the fact that they were not eligible for the benefits of the Orphan Drug Act.

The story of cerezyme, the drug used to treat Gaucher disease, illustrates how the Orphan Drug Act has led to an effective treatment for this disease, but at an exorbitant price.[31] The tale begins with a distinguished physician named Roscoe Brady at the National Institutes of Health. Dr. Brady became interested in metabolic storage diseases and the possibility of correcting them with enzyme therapy. By the early 1970s, his associate, Dr. Peter Pentschev, using human placentas, had managed to isolate tiny amounts of glucocerebrosidase, the defective enzyme in Gaucher disease. Two patients received the enzyme and the initial results looked promising.

A large-scale purification method had been developed by 1977, but the resulting enzyme preparation gave inconsistent results. As we have seen, to be effective, the sugar chains of the enzyme had to end with mannose in order to interact with the mannose receptors on the macrophages. The problem was that the mannose residues on the sugar side chains of glucocerebrosidase were internal in position, so Brady's team succeeded in developing a method for stripping off the blocking sugar molecules to expose the mannose residues. In the first clinical trial involving eight people with Gaucher disease, only the smallest one, a child, showed beneficial effects. The other seven people in the trial were adults and the enzyme replacement therapy did not help them. In the second clinical trial involving 12 individuals, the enzyme dosage was increased and all responded positively. By the end of 1975, the NIH researchers had patented their method for isolating the enzyme.

To carry out the clinical trials necessary to gain FDA approval of enzyme therapy for Gaucher disease, Brady's team at NIH needed a regular supply of enzyme. Hence, the agency contracted with the New England Enzyme Center at Tufts University Medical School to supply enzyme prepared according to NIH procedures. The director of the center was Henry Blair and the NIH contracts over a period of several years amounted to nearly $1 million dollars. In 1981, the center closed and a biotechnology entrepreneur named Sheridan Snyder founded Genzyme together with Henry Blair.

For the next 11 years, the NIH held contracts with Genzyme that amounted to nearly $9 million under which Genzyme provided the NIH with enzyme for clinical trials. Because the NIH had patented its procedure for enzyme isolation and investigators there had published papers on the subject, Genzyme could not obtain a patent for the drug it was now providing to NIH, but in 1985, Genzyme's Alglucerase, as it was then called, received official designation as an orphan drug. This meant that the company would have exclusive marketing rights for Alglucerase for seven years and, hence, enhanced profits. Furthermore, it was unlikely that a competitive product would appear on the market because of the costs of developing another drug for such a small market. Hence, Genzyme would probably be able to sell Alglucerase for a high price without fear of competition after the seven-year period of protection had elapsed. There was a hitch, however. Genzyme could not market its enzyme preparation without FDA approval and this would not be forthcoming until April 5, 1991.

Despite the steady stream of NIH contracts and the money the company made from selling certain chemicals to bigger companies, Genzyme was in trouble within two years of its formation. Headhunters sought out a new CEO and made a pitch to Henri Termeer who had previously been employed by Baxter. Because the Gaucher drug, now renamed Cerezyme, would be the only FDA-approved drug available to treat Gaucher disease, Termeer realized that he could sell Cerezyme to patients for a very high price. Upon FDA approval in 1991, Genzyme brought Cerezyme on the market for an average of $200,000 per patient per year.

Genzyme explained that the price of Cerezyme was so high because it originally took 22,000 placentas to produce enough of the drug to treat a patient for a year. In 1994, Genzyme developed a method for producing Cerezyme from genetically modified cells in culture, but the $200,000 price remained the same. In 2004, about one-third of Genzyme's $2.2 billion in revenue came from the sale of Cerezyme. In 2007, Cerezyme sales amounted to $1.1 billion.

The price of Cerezyme placed it beyond the pocketbooks of most patients, so it was essential that insurance companies pick up at least part of the tab. The problem for Ed and Peggy DeGranier, whose son Brian had Gaucher disease, was that Ed worked for a small employer. Getting the insurer to pay for Brian was going to make all of the company's insurance premiums unaffordable. So Ed DeGranier quit his job and went to work selling cable services door-to-door for Comcast, reasoning that the expense of the insurance payments for Brian's treatment would now be spread among thousands of employees.

Henri Termeer was becoming rich and famous. Massachusetts Governor Deval Patrick appointed Termeer to his council of economic advisers and hosted the Termeers for dinner in the spring of 2008. Termeer himself had earned more than $50 million in total compensation and, in the spring of 2008, was worth around $260 million because of the performance of Genzyme's stock. But then disaster struck. The company experienced serious production problems, not only for Cerezyme, but also for Fabrazyme, a product designed to combat Fabry's disease, another lysosomal storage disease. In June of 2009, Genzyme had to shut its main factory in Boston temporarily due to viral contamination. The company first predicted that production should return to normal in six to eight weeks, but there were repeated setbacks. The value of Genzyme stock began to decline and the production problems opened the company to a proxy challenge by investor Carl Icahn.

In November 2009, Adam Feuerstein, writing for the *Street*, picked Termeer for the 2009 Worst Biotech CEO of the Year Award.[32] Feuerstein blamed Termeer for "fostering an arrogant, irresponsible business culture in which employees are rewarded more for being loyal to the CEO than they are for being competent at their jobs." Feuerstein accused the company of "sloppy manufacturing methods and a dirty, run-down manufacturing plant in Boston that has become an ongoing target for FDA inspectors."

The shortage of Cerezyme allowed two other companies to seek approval of their own Gaucher drugs.[33] Both companies received help from the FDA because the agency had good reason to feel that the sole supplier model no longer worked. In February 2010, one of these companies, Shire, received FDA approval of its Gaucher drug. In September, the EU approved Shire's new drug. Shire planned to sell its drug at a 15% discount to Cerezyme, still expensive, but at last there is price competition for Gaucher drugs.

Big pharmaceutical firms often like to acquire smaller companies like Genzyme that have one or a few highly profitable products. In August 2010, the Sanofi-Aventis offered to buy Genzyme for $69 a share.[34] Genzyme's Board of Directors rejected the offer so the

French pharmaceutical firm launched a hostile takeover in October of that year.

There are several lessons to be learned from the Genzyme story. First, it was because of basic research done by Roscoe Brady and his group at NIH that treatment of lysosomal storage diseases like Gaucher became possible through enzyme therapy. To scale up the process for clinical trials, Brady and his colleagues turned to the New England Enzyme Center at Tufts University run by Henry Blair. Blair and a colleague subsequently founded Genzyme, whose principal money earner became the Gaucher drug Cerezyme. In short, no original research by Genzyme was involved, although they later developed a method for obtaining Cerezyme from tissue culture cells rather than placentas. Because the company benefited from the Orphan Drug Act, Genzyme could charge anything it wanted for the drug once Cerezyme had been approved by the FDA because a competitive drug could not be brought to market for 7 years. In fact, because the number of Gaucher patients was small, Genzyme enjoyed a monopoly on enzyme therapy for Gaucher disease for almost 20 years. The downside of this for patients was that Genzyme charged an outrageous price for Cerezyme and, when the company experienced manufacturing problems in 2009, there was no other place for them to turn to for the drug.

A quite different approach to reducing the amounts of toxic substances that accumulate in lysosomes in storage diseases is called substrate reduction therapy.[35] The idea here is to make use of a small molecule that inhibits or limits synthesis of a precursor to the substrate of the defective enzyme. This treatment has been approved for patients with mild to moderate Type 1 Gaucher's disease.

One final example of enzyme replacement therapy involves a disease called Severe Combined Immunodeficiency or SCID. There are several forms of this disease caused by mutations in different genes, but they all result, as the name suggests, in a highly defective immune system. This means that babies born with the disease are very susceptible to viral, bacterial, and fungal infections and to "opportunistic" organisms that do not normally cause people with functioning immune systems to get sick. Infants with SCID usually develop pneumonia and diarrhea. They do not survive for more than a year or two.

The form of the disease treatable by enzyme replacement results from mutations in the *ADA* gene on chromosome 20. This gene encodes an enzyme called adenosine deaminase.[36] ADA-SCID is very rare, occurring in around 1 in 200,000 to 1,000,000 infants worldwide. Adenosine deaminase is found throughout the body, but assumes critical importance in the immune system. The immune system includes white blood cells called B-cell lymphocytes and T-cell lymphocytes. B cells produce antibodies to invading bacteria and viruses that mark them for destruction. T cells do not produce antibodies, but regulate B-cell activity. Helper T cells assist in the maturation of B cells while regulatory (suppressor) T cells shut down T cell-mediated immunity and, therefore, antibody production at the end of an immune reaction. Cytotoxic T cells destroy virus-infected cells and cancerous cells.

Adenosine deaminase converts a compound that is toxic to T cells (deoxyadenosine) to a harmless compound (deoxyinosine). Mutations in the *ADA* gene that inactivate the enzyme result in a buildup of deoxyadenosine in T cells that result in their death. Obviously, this results in loss of immune system control and is lethal before long.

The therapy of choice for any kind of SCID for over 20 years has been bone marrow/stem cell transplantation from a donor, usually a sibling, who is immunologically compatible (HLA identical). However, enzyme replacement therapy has also been available for ADA-SCID for over 15 years. Treatment of this disease, like the lysosomal storage diseases, is very expensive. Biweekly injections of a modified form of the enzyme are required and the cost may exceed $200,000–$300,000 per patient annually, but the treatment works over the long term.[37] Even though lymphocyte counts are below normal in enzyme-treated patients, their immune systems function well enough to ward off major infections and to foster good health.

Gene therapy: mirage or grail?

Rare genetic diseases come to the attention of the public only occasionally, usually when a newspaper article is published on an otherwise obscure ailment and often in connection with some impending form of therapy. One disease that gained a lot of attention in the

1970s and early 1980s was SCID-X1.[38] This disease is caused by mutations in a gene found on the X chromosome and is almost exclusively found in males. The reason for all the attention was that David Vetter, a little boy with this disease, was encased in a germ-free plastic bubble for ten years to prevent him from contracting an infectious disease.

David's parents, David Joseph Vetter Jr. and Carol Ann Vetter, had a daughter named Katharine. They also had a son named David who had died seven months after his birth from SCID. Three physicians from Baylor College of Medicine, John Montgomery, Mary Ann South, and Raphael Wilson, told the Vetters that another son had a 50% chance of having SCID because Carol Ann carried an X chromosome with the mutant form of the gene. However, they assured the Vetters that, if they had a son with SCID-X1, the baby could be put in a sterile chamber until a bone marrow transplant could be performed. Their assumption was that sister Katharine would be a suitable donor.

In due course, the Vetters had a second son, whom they also named David, and he had SCID-X1. The baby was placed in his sterile incubator pending the bone marrow transplant from his sister. However, his sister was not a good match and a good match was never found. So little David lived in plastic isolation for ten years on the third floor of the West Tower of Texas Children's Hospital in Houston where all he could see was Fannin Street from the room's window. Occasionally, David would be transferred to a bubble at his home for a few weeks. A psychologist named Mary Murphy befriended David, spending a lot time with him. She said that David grew angry. He didn't see the point of learning to read or of schooling. "What good will it do?" he said, "I won't ever be able to do anything anyway."

Over the years, David's three original doctors moved on to positions elsewhere. Two new doctors, Ralph Feigin and William Shearer, took over. Despite exhaustive searching, no suitable bone marrow donor had appeared. The new doctors tried to convince the Vetters to take David home and to try to keep him healthy on a regime of gamma globulin and antibiotics. Perhaps his immune system had improved. The Vetters rejected this solution after having consulted the original trio of doctors.

Another four years passed and by now doctors in Boston had had some success in transplanting unmatched bone marrow. At this point, all of the doctors concurred that using unmatched bone marrow was a risk worth taking. David's sister Katharine donated the bone marrow. It was treated in Boston and flown back to Houston. The transplant was performed on October 21, 1983.

For a few months, everything seemed to go well and the doctors began to hope that David would be able to leave his bubble. But in early February 1984, David became violently ill and he died on February 22. An autopsy revealed that David's body was full of tumors. He had died of a cancer called Burkitt's lymphoma. It turned out that the screening of Katharine's marrow had missed the presence of Epstein-Barr virus, an important contributor to this cancer.

Considering that immunologically matched bone marrow transplants are regarded as the most satisfactory treatments for the different forms of SCID, bone marrow cells of a person with the disease would seem ideal targets for gene therapy. Assuming that a good gene could be introduced into the appropriate cell type and functioned there, the whole immunological matching problem would be obviated. It should be noted that this sort of gene therapy is referred to as somatic gene therapy. The defective gene is repaired only in the targeted somatic cells. This is the only kind of gene therapy possible in humans at the moment. Reproductive gene therapy, where the defect is corrected in eggs and sperm, is not even on the horizon.

W. French Anderson and colleagues conducted the first serious attempt at gene therapy and it involved the *ADA* gene. Getting approval for a human gene therapy trial was a drawn-out process. Anderson's quest for approval began in June 1988.[39] The NIH and the FDA subjected Anderson's clinical protocols to detailed technical review. In early 1990, Anderson cleared the final hurdle and received approval for his gene therapy test from the Recombinant DNA Advisory Committee of the NIH.

And Anderson had a suitable patient. She was an ADA-deficient little girl named Ashanti DeSilva. She was only four and was receiving enzyme replacement therapy. To deliver the gene, Anderson needed an appropriate vector. These vectors are usually disabled viruses. The vector Anderson chose to use to insert a good copy of the *ADA* gene into Ashanti's T cells was a modified retrovirus. These are RNA viruses that copy themselves into double-stranded DNA and in this form insert themselves into the genome (refer to Chapter 1).

There are lots of retroviral remains littering the human genome. The basic structure of a retrovirus is quite simple. There are three genes called *gag*, *pol*, and *env* flanked on either end by long terminal repeats (LTR). Creation of a retroviral vector usually involves removing the three viral genes from a DNA copy of the virus. The therapeutic gene is then inserted in place of them. To prepare virus particles to infect the target cell, the modified retroviral DNA is transferred to a packaging cell line. This cell line has been modified to synthesize the viral proteins required to make virus particles. These particles are then used to deliver the human gene to the target cells that are maintained in culture.

On September 14, 1990, Ashanti deSilva was treated for the first time with an infusion of her own T cells into which a normal *ADA* gene had been inserted. The retroviral vector carrying the normal *ADA* gene had been mixed with T cells that had been extracted from Ashanti's blood and subsequently grown in culture. When these cells were returned by injection to the little girl, they began to produce the ADA enzyme. Because Ashanti seemed to benefit and there were no obvious harmful effects, Anderson and his team repeated the same protocol with nine-year-old Cynthia Cutshall on January 30, 1991. Both girls' immune systems now began to function properly. To be on the safe side, the two girls also continued on enzyme replacement therapy.

Because T cells turn over, injection of genetically modified T cells continued with each girl receiving approximately ten infusions over 2 years. When the girls were monitored after 4 and then 12 years, ADA expression in T cells could still be detected in Ashanti's case.[40] Around 20% of her T cells still carried and expressed the introduced ADA gene. However, in Cynthia Cutter's case, the persistence of cells with the ADA gene was very low (<0.1%) and no expression of the gene was detectable.

Anderson's success with ADA-SCID stimulated a number of proposals for gene therapy trials. By 2000, investigators had initiated over 400 of them. But in 1999, tragedy struck. Dr. James M. Wilson, who had been recruited by the University of Pennsylvania in 1993 to establish an Institute for Human Gene Therapy, was in the process of conducting a liver gene therapy trial. One of his subjects was a young man of 18 from Tucson, Arizona, named Jesse Gelsinger.[41]

Shortly after Wilson arrived at Penn, he had met Dr. Mark Batshaw, an expert on metabolic diseases. Batshaw was particularly interested in a disease that results from a deficiency in an enzyme called ornithine transcarbamylase (OTCD). This enzyme is involved in the synthesis of the amino acid arginine. It is also one of the urea cycle enzymes that convert ammonia to urea. If the cycle is disrupted, as it is in OTCD, ammonia accumulates in the liver, causing liver failure and death.

The *OTC* gene encoding ornithine transcarbamylase is on the X chromosome, meaning that OTCD usually affects males. Newborn baby boys that completely lack the enzyme enter a coma within three days and die if the disease is not treated. Even with current treatment methods, most surviving children have severe cognitive defects.

A crucial step in preparation for the OTCD trial was the development of a vector that would deliver the functional gene to liver hepatocytes. These are the major functional cells of the liver, making up 70%–80% of the cytoplasmic mass. They are also fairly stable with a life span of around five months. The vector chosen was a disabled or attenuated version of an adenovirus that could express the normal *OTC* gene. Adenoviruses are DNA viruses that commonly cause respiratory infections or colds. Adenoviral vectors have the advantage that they infect cells very readily and can accommodate a substantial foreign DNA insert. However, they do not integrate into the genome nor do they or the genes they contain replicate. This meant that, if the procedure worked, it would have to be repeated from time to time as new hepatocytes replaced the older ones.

Dr. Wilson and his colleagues submitted a grant proposal to the NIH on March 23, 1994, and they soon obtained promising data. They found, using a mouse model, that the OTCD metabolic defect could be eliminated by their modified adenovirus for several weeks to a month. High doses of virus were administered both to mice and rhesus monkeys to assess the potential for toxic effects. At the very highest dose, the monkeys developed severe liver disease and a clotting disorder that would have been lethal had the monkeys not been euthanized.

The investigators received permission to enroll subjects in a clinical trial on October 21, 1996, by which time the adenovirus vector they were using had undergone significant improvements. They decided that adults and not infants would be admitted to the trial because only adults were capable of informed consent. Eighteen subjects were enrolled in the clinical trial. Most of the subjects were female carriers of a defective *OTC* gene, but four, including Jesse Gelsinger, were males who had functional and nonfunctional versions of the gene despite the fact that it was located on the X chromosome. The clinical trial was initiated with its first subject on April 7, 1997.

The first 17 subjects enrolled in the trial exhibited a number of surprising side effects and immunological events, some being unpredicted by the preclinical trials. Furthermore, there was no evidence that the introduced *OTC* gene was expressed. Subject 18 was Jesse Gelsinger. He had a partial deficiency in the OTC enzyme that had been diagnosed at 30 months of age following an episode of vomiting, seizures, and coma associated with elevated ammonia levels. Despite dietary and other treatments, he continued to display episodes of distress associated with high levels of ammonia from time to time.

When Gelsinger signed up for the clinical trial, he realized it was to test the efficacy of gene therapy for infants and that it probably would not benefit him. Nevertheless, if it did, he might be released from his restrictive diet where half a hot dog was a treat. As he said to a friend shortly before he left for Penn Hospital in Philadelphia, "What's the worst that can happen to me? I die, and it's for the babies."[42] And that, sad to say, is precisely what happened. When Gelsinger underwent treatment with the adenovirus vector carrying the OTC gene, he developed a massive systemic inflammatory response, multiple organ failure, and died 98 hours after being treated.

Reactions to Gelsinger's death were not long in coming.[43] Officials at the FDA said that Jesse Gelsinger should have been judged ineligible for the trial because his liver was not functioning well enough before he had been infused with the vector. The agency temporarily shut down gene therapy experiments at the University of Pennsylvania. The Senate got involved complaining that gene therapy trials needed proper oversight. A six-member panel led by Dr. William H. Danforth, Chancellor Emeritus of Washington University in St. Louis, recommended, but did not insist, that clinical trials at the Institute for Human Gene Therapy be discontinued. The university implemented this suggestion.

One year and a day after Jesse Gelsinger died, a wrongful death suit was filed in state court in Philadelphia. It accused Dr. Wilson and colleagues of failing to follow federal rules properly and failing to inform Jesse Gelsinger of the risks he faced. Not only were Dr. Wilson and two of his colleagues named in the suit, but Arthur Caplan, Director of the University's Center of Bioethics, was also named as a defendant. In early November 2000, the University of Pennsylvania settled the suit out of court.

Then, just as gene therapy seemed to have entered the slough of despond, an exciting report was published by a French team led by Dr. Alain Fischer, an immunologist and gene therapist, at the Necker Hospital in Paris.[44] They had used retroviral gene therapy to treat SCID-X1 disease in two little boys aged 11 and 8 months at the time of the trial.

SCID-X1 is a very different disease than ADA-SCID. It is the most common form and results from mutations in the *IL2RG* gene located on the X chromosome. The protein encoded by this gene is required for normal immune system function. Without it, T and natural killer (NK) cells are absent and B cells are nonfunctional. A retroviral vector with the functional *IL2RG* gene was used to treat specific bone marrow cells called CD34+. CD34 is a protein marker for stem cells of the blood-forming (hematopoietic) system. By treating these cells, the French investigators were aiming for a long-term solution because these cells give rise to the differentiated cells of the immune system (e.g., B and T lymphocytes).

Then in the spring of 2002, a cute little red-haired boy 18 months in age named Rhys Evans from South Wales was reported to have been cured of SCID-X1 using gene therapy.[45] Dr. Adrian Thrasher at London's Great Ormond Street Hospital for Children led the team that cured Rhys. By Christmas, Rhys' condition had vastly improved, so much so that his parents made a video of Rhys playing under the Christmas tree to send to his doctors as a thank you.

More children were treated, but then the storm clouds gathered again.[46] Four of the ten children treated by the Paris team developed a leukemialike disease between three and six years of age. In the first two Parisian cases, the vector had integrated near a proto-oncogene called *LMO2* and activated it causing cell proliferation. In the third case, there was also vector integration next to *LMO2* and also near another proto-oncogene called *BM11*, but once again it was the event involving *LMO2* that led to leukemia. In the final case, a third proto-oncogene *CCND2* was involved. In three of the four patients, chemotherapy was successful in restoring normal T cell populations, but failed in the fourth patient who died. Then in 2007, the eighth little boy treated by the Thrasher team in London also developed leukemia. Once again, integration leading to the expression of *LMO2* seemed to be the root cause.

A potentially fatal problem now manifested itself.[47] Was the therapeutic gene itself causing cancer? A group led by Inder Verma, an expert on the development of viral vectors for gene therapy, published a paper in *Nature* titled "Therapeutic Gene Causing Lymphoma." Using a mouse model of SCID-X1, they introduced the *IL2RG* gene into the experimental mice and followed them for up to a year and a half. One-third of the mice that underwent gene therapy developed T cell lymphomas. Verma and colleagues concluded that *IL2RG* was itself the culprit. They recommended that preclinical treatments with transgenes should include a long-term follow-up prior to clinical trials.

The teams in Paris and London argued that, for several reasons, the murine results did not apply to the human trials. For one thing, their own experiments with the murine model resulted in only three cases of lymphoma in one trial where 68 mice (26 SCID and 42 normal) had been tested. There were none in two other transgenic lines in which 54 normal mice were tested.

In 2002 in the United States, the FDA put three clinical trials of SCID therapy on hold following a French announcement that one of the boys in their trial had developed leukemia.[48] The trials started once again only to be halted in 2005 after Fischer's team announced early that year that another boy, treated in 2002, had developed leukemia. That same year, a trial of ADA-SCID was allowed to go forward as there was no evidence that this therapy could lead to cancer. In 2010, Children's Hospital in Boston announced that it was initiating a new gene therapy trial for SCID-X1 in collaboration with the Institute of Child Health in London and the Hannover Medical School in Germany. A redesigned retrovirus is to be employed with the likely cancer-causing elements eliminated. It will be used to introduce the *IL2RG* gene.

Meanwhile, the results for ADA-SCID, conducted on patients by investigators at the San Raffaele Telethon Institute for Gene Therapy in Milan, have been even more encouraging.[49] So far, ten patients have been treated. All ten are well. One has been followed up for eight years now and eight are no longer receiving enzyme replacement therapy. There have not been any reported cases of cancer.

Another gene therapy trial that is showing great promise involves vision.[50] During the fall of 2009, the newspapers were full of stories about an eight-year-old boy from Hadley, New York, named Corey Haas. Corey has a rare genetic disease called Leber's congenital amaurosis that was causing him to go blind. Children with this disease start to lose sight at birth and are completely blind by the age of 40. They have a defect in the *RPE65* gene. This gene encodes a protein that is important in the production of a visual pigment called 11-cis retinal and also in the regeneration of visual pigment.

Corey enrolled in a clinical trial carried out by a team at the University of Pennsylvania in collaboration with the Children's Hospital of Philadelphia. His left eye showed the greatest impairment and it was this eye that was injected with a viral vector called adeno-associated virus (AAV) carrying a good copy of the *RPE65* gene. AAV is a small DNA virus that, in contrast to adenoviruses, does not elicit an immune response, but like adenoviruses does not integrate into human genomic DNA. Vision in Corey's left eye improved so rapidly that by the age of nine, he was playing little league baseball, driving go-carts, hiking along trails near Hadley, and was able to read the blackboard in his classroom.

At eight years old, Corey was the youngest of 12 patients in his clinical trial with the oldest being 44. Although all 12 patients showed visual improvement, it was clear that the degree of improvement was age dependent with the youngest patients faring best. The results supported the investigators' hypothesis that the success of gene therapy depends on the extent of retinal degeneration and that increases with age.

As Carolyn Johnson of the *Boston Globe* wrote on February 28, 2009, "Gene therapy started the year with a bang."[51] But it has taken a long time. Between the first proposals for clinical trials over 20 years ago and 2008, over 1,300 such trials have been initiated in 28 countries and the success rate has been very low.[52] Perhaps that will change now with better gene delivery methods, not all of which involve viral vectors, and a better understanding of risk factors. Certainly in the case of a disease like SCID-X1, where enzyme therapy is not an option, the choice is between bone marrow transplant, ideally with a well-matched donor, and gene therapy. And now gene therapy seems to be working well over the long term for ADA-SCID therapy and for Leber's congenital amaurosis.

The great new hope is for stem cell therapy. Not only has President Obama lifted the ban on federally funded embryonic stem cell research, but great progress has been made in creating stem cells from adult cells that gain the potential of being able to differentiate into virtually any cell type. So far, however, only stem cells from hematopoietic tissue have been used successfully in gene therapy in the case of SCID. But myoblasts appear to be good candidates for gene therapy because these muscle stem cells fuse with nearby muscle tissue and become an integral part of muscle.

But James Wilson, in whose clinical trial Jesse Gelsinger died, has reflected on lessons taught for stem cell therapy from gene therapy's rocky road.[53] In the 1990s, the sky seemed to be the limit for gene therapy. It was hyped as the potential cure for a great array of ailments. But in a review written in 2000, the FDA concluded that "the hyperbole has exceeded the results" and "little has worked." Wilson observes, "many of the social and economic forces that drove gene therapy's burst of clinical activity also exist today in the stem cell arena." He points out that the rapid acceleration of gene therapy experiments from bench and model to clinical trials was driven by many factors, including the following: (1) a "simplistic, theoretical model that the approach 'ought to' work"; (2) "a large population of patients with disabling or lethal diseases and their affiliated foundations harboring fervent hopes that this novel therapy could help them"; (3) "unbridled enthusiasm of some scientists in the field, fueled by uncritical media coverage"; and (4) "commercial development by the biotechnology industry during an era in which value and liquidity could be achieved almost entirely on promise, irrespective of actual results."

It would seem wise for stem cell researchers to keep these four points in mind as they forge ahead into this promising new field with its great potential. If gene therapy is any guide, there may be unexpected and, possibly, serious disappointments ahead.

We should soon know more about the potential and pitfalls of stem cell therapy. In January 2009, the FDA approved plans by a company called Geron to conduct the first clinical trial involving embryonic stem cells.[54] These stem cells were manipulated to act as precursors of certain kinds of nerve cells to try to repair recently incurred spinal cord injuries. The first trial is really designed to assess the safety of the procedure. Its first patient was treated in the fall of 2010 at the Shepherd Center in Atlanta, Georgia, a facility that specializes in brain and spinal injuries.

Creating a life to save a life

Fanconi's anemia is a collection of recessive genetic defects affecting any one of 13 different genes.[55] The products of these genes are involved in repairing damage to DNA. In fact, the protein products of eight of these genes form a single repair complex. The major cause of death from Fanconi's anemia is bone marrow failure that usually occurs in childhood.

In 1994, Mark Hughes (refer to Chapter 8, "Preventing genetic disease") got a phone call from John Wagner, an expert transplant surgeon at the University of Minnesota.[56] Wagner often dealt with Fanconi patients, meaning a bone marrow transplant was in order. If a closely matched immune system could be found, for example from a sibling, there was an 85% chance that the patient would survive, but if the match was unrelated, the success rate dropped to 30%. Wagner wondered whether a better solution might be in vitro fertilization (IVF) coupled with preimplantation genetic diagnosis (PGD). Suppose a couple had a child with Fanconi's anemia. If the parents chose to couple IVF with PGD to have another child, that child might provide the right immunological match for the Fanconi child. The ethical problem, of course, was that the couple had to want to have another baby not just to save the Fanconi child, but because they genuinely desired a new baby.

Hughes was concerned for political, but not for technical reasons for he was then at Georgetown, a Catholic school that opposed any sort of fetal research. Eventually, Hughes and Wagner reached a compromise. Hughes would screen embryos of parents who had a child with Fanconi's anemia, but only if the parents were young, carried the most common Fanconi mutation, and had planned to have additional children anyway. The mutation in question is an A-T to T-A change in the *FANCC* gene in the fourth intron. This mutation, particularly common in Ashkenazi Jewish populations, causes a severe form of the disease that takes hold early in childhood. Furthermore, the chances of survival following an unrelated bone marrow transplant for this particular manifestation of the disease are not 30%, but close to zero.

Using these criteria, Hughes and Wagner contacted Arleen Auerbach at Rockefeller University who maintains the national Fanconi registry. They identified two suitable couples for IVF and PGD. They were Lisa and John Nash whose 20-month-old daughter Molly had Fanconi's anemia and Laurie Strongin and Allen Goldberg whose 5-month-old son Henry had the disease.

Auerbach had already told the Nashes and the Strongin-Goldbergs the bad news about the type of Fanconi's anemia their two children had. But she continued that the one thing they had going in their favor was that Henry and Molly were still very young. After deciding with Wagner and Hughes that these two children would be suitable for IVF and PGD, Auerbach called the Nashes in Denver and then the Strongin-Goldbergs in Washington to ask them whether they were willing to make use of these procedures to produce a healthy baby that might be a good immunological match for their Fanconi children.

The Nashes immediately said yes, but Laurie Strongin had just taken a home pregnancy test and found she was expecting. Of course, if the baby to come was healthy and a good match for Henry there would be no problem. However, when Jack Strongin-Goldberg was born in December 1995, he proved to be a poor immunological match, even though he was free of Fanconi's anemia. So, the Strongin-Goldbergs were going to have to have yet another baby if they were to save Henry.

Just about the time that Jack was being born, Lisa Nash was getting shots to help her produce eggs for in vitro fertilization, but following the procedure, she failed to get pregnant. At the same time, Mark Hughes got into trouble with NIH and his wife became seriously ill with breast cancer. He was accused of using federal funds for embryo research, a violation of the Congressional ban. Hughes had scrupulously kept his federally funded research completely separate from his attempts to aid the two Fanconi families. Furthermore, he argued that he was not working with embryos at all. He was only characterizing the DNA from a biopsied embryonic cell that had been sent to him for this purpose. His arguments were rejected. In the end, Mark Hughes gave up his grant and left Georgetown University for a privately funded institute associated with Wayne State University near Detroit where neither federal nor ecclesiastical meddling would obstruct his work.

The Strongin-Goldbergs and the Nashes finally met in the spring of 1997. Laurie and Lisa soon became good friends, telephoning frequently to compare notes on their ailing offspring. In January 1998, Hughes was finally ready to go forward again in his Wayne State laboratory. Laurie took the train to New York and underwent in vitro fertilization at Weill Cornell Medical Center under the direction of Dr. Zev Rosenwaks, an expert in the field. Her husband flew the 16 fertilized embryos to Detroit where Hughes picked them up. Two were immunologically perfect matches for Henry, but both had Fanconi's anemia.

In the end, Laurie Strongin and Allen Goldberg would undergo nine unsuccessful attempts to create a sibling to save Henry. During her ordeal, Laurie was injected 353 times and produced 198 eggs.

Meanwhile, Henry's blood counts were dropping. On June 6, 2000, Henry received a bone marrow transplant from an anonymous donor. It was not a success. Henry got worse and died in early 2003.

The Nash family was more fortunate. Molly's blood counts kept dropping and they kept sending Mark Hughes frantic e-mail messages. In the summer of 1998, he informed them that he could do no more for them. Hughes told them that he was unable to access critical records in the possession of his former lab at Georgetown because the university administration refused to get involved in any way. Then they learned, almost by accident, about Yuri Verlinsky's private clinic in Chicago (refer to Chapter 8). There they met Charles Strom, who was then head of the Reproductive Genetics Institute's genetics lab. He agreed to try to help the Nashes. For much of the time the story progressed as it had for the Strongin-Goldbergs. There were matches, miscarriages, and failed pregnancies. But then the Nashes approached Dr. William Schoolcraft, a fertility doctor in their native Colorado. Lisa had been producing relatively few eggs per cycle so Schoolcraft insisted on a procedural change to Lisa's hormone therapy to boost her egg production.

In December 1999, Strom retrieved 24 eggs from Lisa and, after IVF, reported that there was a single match. Lisa became pregnant, but it was a very precarious pregnancy punctuated by frightening bleeding episodes. In August 1999, Lisa gave birth to a little boy, Adam. His cord blood was kept and it saved Molly Nash's life. In March 2010, Laurie Strongin published *Saving Henry: A Mother's Journey*. It was a difficult journey indeed and one rarely taken.

Concluding thoughts

As Dr. Kenneth Pass noted when discussing New York's newborn-screening program, although genetic diseases individually are rare, they are not so rare collectively. Genetic diseases, unlike most other diseases, are forever and they often get worse with time. Each one demands a unique kind of treatment that can often be very costly. For tales like the one told by Laurie Strongin about her son Henry, there are thousands of other equally heart-rending stories that never get told at all. Pedigrees can sometimes help to alert prospective parents to a potential problem, but there are lots of instances where the disease is rare enough that carriers who marry and have children are often unaware that one or more of their children may be victims.

As the last chapter of this book unfolds, the scene is changing. Genomic sequencing is getting cheaper and faster. Before long, hidden disease genes will be revealed to prospective parents who want to know their genetic bills of health before they have children. And often enough, a couple will find they do carry mutant genes for some obscure genetic disease. They will then have to decide how to proceed. Should they take the one in four odds for a genetically recessive mutation, for instance, that they will have a defective child and go forward the natural way? Maybe the prospective mother can have chorionic villus sampling or amniocentesis to detect the defect in the fetus. But should the fetus be genetically diseased, the couple will have to decide whether to have an abortion and start all over again.

Alternatively, the couple may choose to have PGD and have only embryos lacking the defect, or perhaps carrier embryos, implanted. This itself is not necessarily a fail-safe procedure and it becomes less so as the prospective mother ages. Furthermore, it is costly and several cycles of IVF may be required and success is not necessarily guaranteed.

These are the kinds of decisions couples will face in the future. They are decisions that will depend both on religious preference and also on how one defines a defect. And with the latter question, we have found our way to the slippery slope. At what point along the spectrum of disease does a couple decide what constitutes an imperfection serious enough to warrant avoidance?

10

The dawn of personalized medicine

Deciphering of the human genome coupled with the HapMap and the ever-decreasing cost of genomic sequencing has brought us to a new era where remedies can be targeted evermore precisely against diseases and, perhaps more disturbingly, the onset of specific diseases can be predicted with more certainty.

None of the genetic disease treatments described in Chapter 9, "Treating genetic disease," is targeted at specific mutations in the defective gene. Instead, they are meant to alleviate the effects of the disease for any mutation in that gene. Dietary prescriptions, for example for phenylketoneuria, apply to anyone with the disease, irrespective of which of the 500 different identified mutations they carry in this gene. The same is true for enzyme replacement therapy and for gene therapy. But in the case of cystic fibrosis, none of these approaches has succeeded. So, a company named Vertex has been working with the Cystic Fibrosis Foundation for a decade to develop products that target the defective CFTR protein produced by specific mutants to try to restore partial function.

One of the products that has promise is named VX-770.[1] This product has undergone clinical trials with patients having a mutation called G551D. This mutation is responsible for about 4% of all cystic fibrosis cases. VX-770 is known as a potentiator, meaning that it opens the closed channel in the CFTR protein made by the mutant. In 2008, Chrissy Falletti, a young Ohio woman, was enrolled in a clinical trial of VX-770.[2] Since childhood, she had faced a daily regime that included 15 different medicines and treatments. Despite this, Falletti competed in state and national competitions in gymnastics when she was only 16. She married an optician and became a first-grade school teacher. By 2007, her lung function was only 50% of normal and she was approaching the age when cystic fibrosis is often lethal.

The clinical trial in which Chrissy Falletti participated was a second, or phase II trial. Normally, a drug is administered to an ever-expanding population of patients in three separate clinical trials. Assuming all three clinical trials show that the drug is efficacious, the drug maker is now in a position to propose to the FDA that it be licensed. Chrissy Falletti took VX-770 orally. It was designed to make the abnormal protein functional so that it could properly facilitate chloride movement. It seemed to work as Falletti said, "On the Vertex study, I just felt completely different. My sister said she used to know when I entered church because she could hear my cough. Then, when I was on the study, my nephew told me, 'I can't tell when you come in anymore.'"

After a little less than a month, Falletti's lung function had improved by 18%. But then the trial ended and Falletti's lung function began to decline once again. She keeps calling the Cystic Fibrosis Center in hopes that VX-770 will be approved for use for her variety of cystic fibrosis before much longer. The expected completion date for the VX-770 clinical trials is 2011.

Another Vertex compound undergoing clinical trials is VX-809.[3] This is called a "corrector" molecule. Its target is the mutant CFTR protein produced in people with the ΔF508 mutation. This is, of course, by far, the most common cause of cystic fibrosis. The deletion of the amino acid phenylalanine at position 508 in the CFTR protein prevents the mutant protein from folding properly and impedes its movement to the membrane surface where it would normally function. The hope is that VX-809 can overcome these problems. So far, the results from clinical trials are encouraging. The next step will be to test VX-809 in combination with VX-770. This will combine a corrector molecule with a potentiator molecule. Several other promising "corrector" molecules have been discovered, but they have not yet been tested in clinical trials.[4]

PTC Therapeutics has taken a different approach that may have wider application.[5] Nonsense mutations within a gene cause premature polypeptide chain termination (refer to Chapter 2, "How genetic diseases arise"). They do this because they are read as the stop codons

(UGA, UAA, UAG) that are normally found at the end of the genetic sentence specifying a protein. PTC Therapeutics sought a compound that would permit an internal nonsense codon to be ignored while allowing stop codon recognition at the end of the genetic sentence at the same time. UGA was the codon of choice and PTC124 emerged as the compound of choice after around 800,000 low molecular weight chemicals had been tested.

Nonsense mutations account for 10% of all CF cases, but in Israel, these mutations are responsible for most CF cases. PTC124, now called ataluren, is proving highly effective in clinical trials not only for CF UGA nonsense mutations, but also for similar mutations in the *DMD* gene encoding the muscle protein dystrophin. These mutations result in Duchenne/Becker muscular dystrophy.

The *DMD* gene is on the X chromosome so males are the predominant victims of Duchenne muscular dystrophy. As we have seen earlier, this is also true in the case of hemophilias A and B that are caused by mutations in the *F8* and *F9* genes, respectively. Males with nonsense mutations in these genes are also participating in clinical trials of ataluren. So this compound is in one sense a broad-spectrum drug in that it may be able to override mutations in many different genes, but ataluren is, at the same time, highly specific in that it is only effective against individuals having UGA nonsense mutations. Like the CF drugs made by Vertex, ataluren does not actually repair the genetic defect that causes the disease, so it must be taken orally on a regular basis.

Pharmacogenomics: tailoring drugs to people

Making drugs that attempt to render mutant proteins functional is just one aspect of personalized medicine. It is equally important to discern whether an existing drug actually benefits or perhaps even harms a person. This can depend on genotype. Unanticipated drug responses are estimated to result in 2 million hospitalizations and 100,000 deaths in the United States each year. By being able to predict such adverse reactions in advance, pharmacogenetics could reduce their number substantially.[6]

About 10% of the labels for drugs approved by the FDA contain pharmacogenomic information.[7] The FDA maintains an extensive list of genetic "biomarkers" for specific drugs that it updates periodically. The list also indicates whether pharmacogenomic testing is required, recommended, or for information only.

For example, testing is required before administering Selzentry (Pfizer). It is prescribed for treating a specific kind of HIV infection.[8] The AIDS virus enters T cells, macrophages, and so on through a cell surface protein complex that includes CD4 and one of two coreceptor proteins, CCR5 or CXCR4. Most primary HIV strains use the CD4-CCR5 complex, and Selzentry blocks entry of the virus through this receptor complex. However, this is not the case for the CD4-CXCR4 complex, so before Selzentry is prescribed, a test is required to determine which entry pathway the virus is using.

Warfarin, perhaps better known by the brand name Coumadin, is a widely used anticoagulant that was initially employed as a rat poison.[9] It has a narrow therapeutic index, meaning that use of too little of the compound can result in stroke or other complications, whereas excess warfarin usage can cause bleeding and hemorrhage. During 2006, more than 36,000 patients with warfarin-related dosage problems were admitted to emergency rooms in the United States.

Blood coagulation involves a cascade of secondary factors that lead ultimately to the formation of fibrin from fibrinogen and to blood clot formation. These, of course, include clotting factors VIII and IX encoded by the *F8* and *F9* genes mutations, which are responsible for hemophilias A and B, respectively (refer to Chapters 1 and 9, "Hunting for disease genes" and "Treating genetic disease," respectively). The activation of several coagulation factors so that they can do their jobs requires vitamin K in the reduced form. That is, it has gained electrons from hydrogen. Warfarin blocks the reduction of vitamin K from its oxidized form, meaning that the clotting factors requiring reduced vitamin K cannot be activated. This blocks clot formation and promotes bleeding.

Warfarin prevents the reduction of vitamin K by interfering with the action of the enzyme that catalyzes this process, vitamin K epoxide reductase (VKOR). The *VKORC1* gene encodes this enzyme and variations within this gene affecting the activity of the enzyme also determine the optimal warfarin dosage. One polymorphism upstream of the *VKORC1* gene in the promoter region results in decreased enzyme production, meaning that reduced warfarin dosages must be administered. This variant is common in Asian and Caucasian populations.

The inactivation of warfarin depends on the product of a gene called *CYP2C9*. It encodes a member of the P450 family of enzymes called Cytochrome P450 2C9 (abbreviated CYP2C9). Enzymes of the P450 family catalyze many reactions related to drug metabolism and are also important in the synthesis of cholesterol, steroids, and other lipids. CYP2C9 metabolizes over 100 therapeutic compounds, including warfarin. There are three variants of the *CYP2C9* gene that, for the sake of simplicity, are referred to as *1, *2, and *3. Individuals having two copies of *CYP2C9* variant 1 (referred to as *CYP2C9*1*1*) can tolerate the highest levels of warfarin, whereas those with two copies of variant 3 (*CYP2C9*3*3*) are the most warfarin-sensitive. The other combinations (i.e., **1*2, *1*3, *2*2, *2*3*) exhibit increasing sensitivity to warfarin in the order shown. Caucasians show considerably more variability in the *CYP2C9* gene than people of Asian or African descent. They are predominantly carriers of the **1*1* version of the gene.

Because of the obvious importance of variations in the *VKORC1* and *CYP2C9* genes in determining proper warfarin dosage, both genes are included in the FDA biomarker list and a company named Genelex offers genetic testing for variants in both genes.[10] Genelex has constructed a matrix of optimal warfarin dosages for different genotypic combinations of the two genes. This matrix is the same as the familiar Punnett Square used to show the different possible combinations of two genes (see Table 10–1).

Another drug for which testing is mandatory is herceptin.[11] This drug is used to counter an aggressive form of breast cancer in which there is overexpression of an oncogene called *HER2*. The name stands for *H*uman *E*pidermal growth factor *R*eceptor 2. This phenomenon occurs in around 25% to 30% of breast cancers. It results either because the *HER2* gene has undergone amplification or the existing copies of the genes are themselves overexpressed. This overexpression can be detected at the protein level using immunological techniques or a method called FISH (fluorescence in situ hybridization) that indicates whether the *HER2* gene has been amplified. Herceptin is a monoclonal antibody against the HER2 receptor protein and targets tumor cells overexpressing this protein for destruction by the immune system. Herceptin binding to the HER2 protein also results in the increase of a protein called p27 that inhibits cell proliferation. Thus, herceptin should only be used to treat those breast cancers where *HER2* is overexpressed.

Table 10–1 Determining proper warfarin dosages based on CYP2C9 and VKORC1 genotypes

VKORC1	***CYP2C9***					
	***1*1**	***1*2**	***1*3**	***2*2**	***2*3**	***3*3**
GG[!]	5–7	5–7	3–4	3–4	3–4	0.5–2
AG	5–7	3–4	3–4	3–4	0.5–2	0.5–2
AA	3–4	3–4	0.5–2	0.5–2	0.5–2	0.5–2

Modified from a table presented by Genelex.
http://www.healthanddna.com/SampleWarfarinReport.pdf
The numbers in the table are recommended dosages of Warfarin in milligrams.

[!]These three different variants 1639 base pairs upstream of the VKORC1 gene affect sensitivity to warfarin dose.

One final example of the importance of pharmacogenomics involves a drug called Elitek.[12] This drug is particularly useful for patients who exhibit tumor lysis syndrome. This occurs in cancer when the malignant cells break open and release their contents in great quantities. The most common problem associated with tumor lysis syndrome is hyperuricemia. This happens because the products released by broken cancer cells include purines derived from DNA and RNA that, during their breakdown, undergo conversion to uric acid. This compound is soluble under physiological conditions (pH 7.4), but crystallizes out of solution under the acidic conditions (pH ca. 5.0) found in the renal tubules and collecting ducts of the kidney. Crystals of uric acid may also get deposited in the joints.

Elitek is a recombinant enzyme that converts uric acid to allantoin. The latter compound is soluble in the kidneys and is eliminated in hyperuricemia patients. The problem is that the reaction leading to allantoin formation also generates hydrogen peroxide. Although this is not a concern for most people, it is a serious complication for individuals with glucose-6-phosphate dehydrogenase deficiency (G6PD), an X-linked hereditary disorder. G6PD deficiency affects 400 million people worldwide. Its high frequency results because it protects against malaria (refer to Chapter 3, "Ethnicity and genetic disease"). The G6PD enzyme plays an essential role in protecting red blood cells against strong oxidizing agents like hydrogen peroxide. In G6PD-deficient patients, treatment with Elitek causes dangerous peroxides to accumulate in red blood cells and massive hemolytic destruction of these cells occurs. Hence, before Elitek is recommended for hyperuricemia treatment, it is very important to determine whether a patient carries a mutation conferring G6PD deficiency.

The $1,000 human genome

In 1953, Francis Crick and James Watson famously published their solution for the structure of DNA in the journal *Nature.*[13] They had come to focus on this molecule as the likely genetic material because of three important observations. In 1944, Oswald Avery, Colin MacLeod, and Maclyn McCarty, working at what was then called the Rockefeller Institute of Medical Research in New York City, published a paper that provided unequivocal evidence that DNA could transmit genetic information.[14] They were experimenting with the pneumonia-causing bacterium *Streptococcus pneumoniae* (Pneumococcus). Cells of virulent forms of Pneumococcus are coated with a polysaccharide capsule and colonies of these forms appear smooth when grown in petri dishes. However, mutants could be isolated that lacked this capsule. These were not virulent and formed rough-looking colonies. Avery, MacLeod, and McCarty first demonstrated that heat-killed smooth coated bacteria could transform rough cells to the smooth condition. Following careful cellular fractionation, they were able to show that the "transforming principle" was DNA and not protein, RNA, or some other cellular component.

A few years later in 1952, Alfred Hershey and Martha Chase, working at the Cold Spring Harbor Laboratories, in Cold Spring Harbor, New York, made a second unequivocal demonstration that DNA must be the genetic material.[15] They were experimenting with bacterial viruses (bacteriophage) that infect *Escherichia coli,* a familiar inhabitant of our gut and a biological workhorse. The virus particles contained DNA wrapped up in a protein coat. Using radioactive labeling, Hershey and Chase demonstrated that only the bacteriophage DNA entered infected cells where it replicated and produced new progeny viruses. And also, in the early 1950s, a biochemist named Erwin Chargaff at Columbia University demonstrated that the four bases in DNA (A, C, G, and T) were present in very specific ratios such that A=T and G=C.[16]

So Watson and Crick knew that DNA must be the genetic material and their model had to account for the equivalence of A with T and G with C. There were other crucial pieces of information. Watson and Crick were working in the Cavendish Laboratories at the University of Cambridge, but they were also in close contact with Maurice Wilkins at King's College in London. Rosalind Franklin, a colleague of Wilkins working with Raymond Gosling, had produced X-ray diffraction images of DNA. These not only provided evidence for the probable helical structure of DNA, but also that the phosphate backbones of the two helices were on the outside of the molecule, meaning that the paired bases must be on the inside. It was Franklin who told Watson and Crick that the backbones had to be on the outside.

In their paper on the structure of DNA, Watson and Crick used some of Franklin's unpublished data. Consequently, many feel that she should have been an author on that paper. In fact, she, herself, was close to solving the helix. In any event, she died of cancer in 1958, four years before Watson, Crick, and Wilkins were awarded the Nobel Prize in Physiology or Medicine.

Publication of the structure of DNA in 1953 initiated an enormously productive era in molecular biology. Soon, the mechanism by which information encoded in DNA was converted to the amino acid sequence of a protein was being unraveled while the genetic code itself was being worked out. Gene regulation became an area of intense interest and by the early 1970s Walter Gilbert and Allan Maxam at Harvard University and Frederick Sanger at the University of Cambridge had developed methods for sequencing DNA.[17] In subsequent years, Sanger's method has prevailed.

By 1988, gene sequencing had progressed to the point that the National Institutes of Health approved the human genome-sequencing project under the direction of James Watson. In 1991, President George H. W. Bush appointed Dr. Bernadine Healy as the Director of NIH. Healy was in favor of patenting genes despite the objections of Watson, so she approved patents on 347 human genes believing that this would promote rather than hinder access to information about them. Watson disagreed and resigned in 1992 to return to Cold Spring Harbor Laboratories as president while Francis Collins took over as head of the Human Genome Project.

A working draft sequence of the human genome was published by the genome consortium in 2000 and a complete sequence in 2003, just 50 years after the publication of the original paper describing the structure of DNA by Watson and Crick. Craig Venter was in hot pursuit (refer to Chapter 2). The public project had cost an estimated $2.7 billion, taken 13 years, and involved over 2,800 scientists at institutions in six different countries (see Table 10–2).[18] Furthermore, the consortium sequence did not represent a single individual, but was a composite sequence that made use of DNA from volunteers who represented diverse populations.

Table 10–2 The first three human genomes*

Genome sequenced	HGP (2003)	Venter (2007)	Watson (2008)
Time taken	13 years	4 years	2 months
Cost	$2.7 billion	$100 million	<$1.5 million
Institutes involved	16	5	2
Countries involved	6	3	1

*Adapted from Wadman, Meredith. "James Watson's Genome Sequenced at High Speed." *Nature* 452 (2008): 788. Note that the Human Genome Project (HGP) sequence is a composite from different individuals.

In 2007, Venter's faster method produced the sequence of his own genome at a cost of around $100 million (see Table 10–2). It employed just 31 scientists in three countries. In 2008, a Branford, Connecticut, company called 454 Life Sciences, acquired by Roche, produced the complete sequence of Jim Watson's genome for less than $1.5 million in a couple of months with a team of 27 scientists in one country, the United States. At the inaugural Consumer Genetics

Show in Boston in June 2009, Jay Flatley, the CEO of San Diego–based Illumina, announced the launch of a personal genome-sequencing service priced at $48,000.[19] A year later, Illumina's price had dropped to under $20,000. Flatley disclosed that the company had so far sequenced the genomes of at least 14 different individuals, including Flatley himself, Henry "Skip" Gates, the Harvard professor, and his father as well as Glenn Close.

Illumina was in close competition with Applied Biosystems and the price kept dropping. As 2010 wore on, Illumina claimed it could sequence a genome for $10,000 and Applied Biosystems countered with a price of $6,000 that then dropped to $3,000.[20]

Another company competing in the cheap genome-sequencing contest is Complete Genomics.[21] The company reported sequencing three human genomes for an average cost of $4,400. It's important to note that these cost estimates are somewhat deceptive. They are usually based on the reported costs of consumable supplies used in sequencing. They do not include the initial investment in sequencing equipment and labor. For example, on the *Science* paper that resulted from Complete Genomics sequencing of three human genomes at an average cost of $4,400, there were 65 authors.[22] This suggests that labor expenses must have been considerable. However, this is of no concern to the consumer as long as the company can afford it. The bottom line is that at $1,000 per genome, a lot of people are likely to have their genomes sequenced.

The technique used by 454 Life Sciences, Illumina, and other companies is called "massively parallel sequencing."[23] Different companies use somewhat distinct technical approaches, but they all involve reading relatively short DNA sequences, the use of specific algorithms to align the sequences, and powerful computers. Faster and better sequencing machines continue to emerge. At the Advances in Genome Biology and Technology meeting held on Marco Island in Florida in February 2010, Hugh Martin, CEO of Pacific Biosciences, described what he calls the first third-generation sequencing machine.[24] It's big, weighs 1,900 pounds, and costs $695,000, but with machines like this, the speed and cost of genomic sequencing will continue to drop.

The X Prize Foundation will award the Archon X Genomics X Prize for the first team to successfully sequence 100 human genomes in

10 days.[25] A plausible contender for this prize is George Church of Harvard Medical School. Church, an established leader in the development of DNA sequencing technology, started up a not-for-profit called Personal Genome Project in 2006.[26] He laid out his goal in a *Scientific American* article titled "Genomes for All." He plus nine volunteers would have their genomes sequenced following which all of that information would be made public. Church's eventual aim would be to sequence the genomes of 100,000 individuals. This would contribute greatly in our understanding of human genetic variation and its relationship to disease. Church started very cleverly by recruiting well-known people like Harvard psychologist and author Steven Pinker, prominent health-care investor Esther Dyson, and Helicos Biosciences founder Stanley Lapidus as guinea pigs. Church's Personal Genome Project has already enrolled 16,000 volunteers.

Measuring genetic risk

In November 2007, deCODE genetics became the first company to offer a direct-to-consumer (DTC) genetic-testing service.[27] The company named it deCODEme. Its purpose was to identify genetic risk factors for certain diseases, including type 2 diabetes and certain cancers. Jeffrey Gulcher, the cofounder of deCODE, submitted a sample of his DNA for analysis. Because he was already getting bald, Gulcher was not surprised that analysis of his genome suggested a propensity toward baldness. Neither was he concerned that he had average risks for a dozen other medical conditions. However, he had twice the average risk of becoming a type 2 diabetic and a one in three chance of developing prostate cancer. Although Gulcher could counter the potential of contracting diabetes with diet and exercise, it was not clear what he should do about his increased chances of suffering from prostate cancer.

Although Gulcher's prostate specific antigen (PSA) tested only a little above normal, his doctor suggested that he consult with Ian Catalona, a highly regarded urologist and prostate cancer specialist, at Northwestern University. Catalona carried out an ultrasound biopsy on Gulcher's prostate that indicated he had a Gleason grade of 6. Gleason grades run from 1 to 5 with 1 being almost normal and 5 being very abnormal. Because two potentially cancerous areas of the prostate are scored to obtain the Gleason score, a very moderate cancer would be a 2(1+1), whereas a very serious cancer would be a 10(5+5), so a score of 6(3+3) meant Gulcher had a mildly aggressive form of prostate cancer. Nevertheless, Catalona removed Gulcher's tumor. It had been detected early and there were no signs of metastasis. So, the deCODE test may have saved Gulcher's life; Kari Steffanson, deCODE's then CEO, certainly thought so.

deCODE is just one of three major DNA testing companies. The other two are Navigenics, founded by David Agus and Dietrich Stephan, and 23andme, whose founders are Linda Avey and Ann Wojcicki, the wife of Google cofounder Sergei Brin. All three companies offer online services that allow customers to order tests for a variety of different genetic traits and conditions with a credit card. Navigenics also offers genetic counseling to its customers to help them interpret the results they receive.

A major concern is how the public will make use of such information in the absence of the opinion of a genetic specialist such as a counselor or doctor. For example, 23andme offers tests for several genetic variants that increase the risk of breast cancer somewhat. However, the company does not test for *BRCA1* and *BRCA2* mutations, although they greatly increase the risk of breast cancer, because Myriad Genetics has the patents for these genes (refer to Chapter 5, "Genes and cancer"). 23andme does point out that if a woman has a familial risk of breast cancer, she should probably consult a doctor "who may suggest getting a clinical genetic test. The effects of rare but serious mutations in genes such as *BRCA1* and *BRCA2* far outweigh those of the SNPs reported here."[28]

A different problem concerns the weighting of genetic risk. This is going to change for a given disease as new molecular markers are discovered. Eventually it seems likely that this problem will be obviated when genetic-testing companies move to whole genome sequencing, as opposed to providing tests for very small sections of the human genome. A company called Knome, cofounded by George Church in April 2009, points to the direction that testing will probably take in the future.[29] The company "provides whole genome sequencing services and interpretation solutions to biomedical researchers and physician-directed families seeking to understand the genetic underpinnings of human disease." Ultimately, it can only be through whole genome sequencing that total genetic risk can be estimated.

So how do you estimate genetic risk anyway? There are two components: relative risk and absolute risk.[30] The average for relative risk is 1.0. So, if you have a risk of 0.5, that means your relative risk is half the normal average, but if it is 1.5, your risk of having the disease or condition is 50% greater than normal. That might seem like a lot, but now let's consider absolute risk. For type 2 diabetes, the absolute risk of contracting the disease before age 60 is 23%. This means that, if your own relative risk is 1.5, the chance that you will become a type 2 diabetic by age 60 is 1.5 x 23 or 34.5%. Conversely, if your relative risk is 0.5, your risk drops to 0.5 x 23 or 11.5%. Obviously, there is a large environmental component involved with this disease as well that needs to be factored into consideration.

At the moment, state regulation of DTC genetic testing is highly variable. As of 2007, 25 states and the District of Columbia allowed unrestricted DTC testing, 13 states prohibited the practice, and 12 permitted testing for specified categories of conditions, but these tended to exclude genetic tests.[31] Both the American College of Medical Genetics and the American Society of Human Genetics have issued cautionary statements about DTC testing.[32] The recommendations of the American College of Medical Genetics stress that, "a knowledgeable professional should be involved in the process of ordering and interpreting a genetic test;" that "the consumer should be fully informed regarding what the test can and cannot say about his or her health;" "that the scientific evidence on which a test is based should be clearly stated;" and that "the clinical testing laboratory must be accredited by the CLIA, the State and/or other applicable accrediting agencies."

The CLIA refers to the Clinical Laboratory Improvement Amendments. These are federal regulations governing proficiency testing of clinical laboratories and the like. By 2011, the NIH expects to have a Genetic Testing Registry in place that will "contain information voluntarily submitted by genetic test providers."[33] At present, it is unclear whether this registry anticipates future federal regulation of DTC testing.

An obvious concern for people contemplating genetic testing, DTC or otherwise, is whether they will be covered by their health insurance or discriminated against in the workplace if a deleterious trait is discovered. The answer is probably not. On May 21, 2008, President George W. Bush signed the Genetic Information Nondiscrimination Act (GINA) into law.[34] The act is designed to prohibit discrimination against individuals by insurance companies and employers based on their genetic information. The purpose of this law is to allow people "to take full advantage of the promise of personalized medicine without fear of discrimination." There are already many state statutes in place that are designed to limit genetic discrimination, but they are highly variable. Some are more protective than GINA and some less so. At a minimum, all state provisions must now comply with GINA.

So from both the scientific and legal point of view, United States citizens will be increasingly able to access their genetic information without fear of discrimination. Most people will be at a loss when trying to weigh the significance of what they learn, partly because they face information overload and partly because they won't be able to understand the meaning of various risk factors or susceptibilities. Some will find they are faced with an uncertain future. Perhaps they are slated to suffer from Parkinson's disease or early-onset Alzheimer's. Hence, it is of paramount importance that people who gain information about what is in their genomes also be counseled by genetic specialists, either counselors or doctors, on what it all means.

Postscript: a cautionary note

In the year 1902, Archibald Garrod made the first connection between a gene and a disease, alkaptoneuria. Today, over 6,000 single gene disorders are known and many genetic risk factors for complex diseases like type 2 diabetes and coronary disease and behavioral conditions like bipolar disease and schizophrenia have been identified. Direct-to-consumer gene-testing companies already provide information to potential disease gene carriers while also offering to assess genetic risk that a person will be susceptible to a complex condition like coronary disease. Even more personal genetic information will soon be on the way as genomic sequencing becomes so cheap that many people will decide to have their sequences read. Francis Collins even predicts that newborn screening for genetic diseases will expand to genomic sequencing of newborns. What this means is that a deluge of personal genetic information is soon going to be spilling out of computers. This raises two important questions. First, what safeguards will there be to make sure this genetic information is kept confidential? Second, how will the average person make sense of the sequence of his or her genome?

Safeguarding of genetic information is going to be of paramount importance particularly if genomic sequencing becomes routine for newborns. We are already familiar with the problems that arise when an insurer or a bank loses a laptop with a raft of personal information. What measures will be taken to ensure this does not happen?

Where genomic sequencing is likely to be particularly important is in the identification of unsuspecting carriers of rare genetic diseases who intend to have children, but for whom there is no familial record of the disease. Suppose that genomic sequencing reveals that a couple intending to have children are both carriers of Type 1 Gaucher disease. Without sequencing, they could have gone ahead and had a baby only to discover later on that their child has Type 1 Gaucher disease. Currently, treating the disease would cost them around $200,000 per year for enzyme therapy with no end in sight. But if genomic sequencing reveals they both carry the Gaucher mutation, they have a couple of other options open to them. The couple may choose to avoid having a Gaucher baby by making use of in vitro fertilization coupled with preimplantation genetic diagnosis. This might cost on average $15,000–$20,000 for a single cycle of in vitro fertilization, including preimplantation genetic diagnosis. This is a lot less costly than enzyme therapy, but success is not guaranteed. In fact, for women under the age of 35, the success rate of in vitro fertilization is 46% and drops with increasing maternal age. Hence, more than one cycle may be required for pregnancy to occur. Then, there is the option of taking your chances, checking the fetus for Type 1 Gaucher disease using chorionic villus sampling or amniocentesis and aborting the fetus if it has Type 1 Gaucher disease. This might cost $1,500, including the physician's fee. A first trimester abortion might cost between $350 and $550. So this is probably the least expensive option, but the couple will have to do it all over again if they abort a Gaucher fetus with no guarantee that the next pregnancy will not yield another fetus with Gaucher. Whether either of these avenues is open to the couple will depend upon how they regard abortion and embryo creation and destruction.

Genomic analysis will undoubtedly turn up all sorts of genetic risk factors and people will need expert advice, either from a genetic counselor or a physician trained in genetics, in order to make sense of this information, some of which they might prefer not to know. It is easy to envision a time not too far away when a large expansion in the number of genetic counselors is going to be needed, especially ones that are familiar with the relationship between genetic risk factors and complex diseases or conditions.

We are all aware of the potential that embryonic and now adult stem cells may have for alleviating specific diseases. For a person with a debilitating genetic disease, the idea of being treated with stem cells from normal embryos produced by couples with no sign of the disease may be very appealing. These cells are pluripotent, meaning they can differentiate into any cell type in the body. Imagine, for instance, that a boy with Duchenne/Becker muscular dystrophy has the opportunity of being treated with embryonic cells that have the potential of integrating into his withering muscles to provide normal muscle function. It's a captivating notion. But wait a minute. We don't know if immune system rejection will be a problem. We don't yet have the clinical trial data that will inform the medical establishment about whether such a treatment is going to work. We have the hype just as we did 20 years ago with gene therapy. Let's just hope the outcome this time is more promising and does not deliver so many disappointments.

Suppose immune rejection is a problem. There are always adult stem cells from the patient that can be used. This should obviate the rejection problem, but at the same time a new problem arises. These mutant cells must be treated with copies of the normal gene. In other words, they must undergo gene therapy just like the stem cells of the hematopoetic system in the case of severe combined immune deficiency disease. The experience with this type of gene therapy is promising so we can only hope that it will work with adult stem cells. Once again, only time will tell what the outcome will be.

In the next few years, there is bound to be lots of news on the cancer front too. Excitement will wax and then wane concerning the promise of some new drugs, but others will prove their worth. It seems very likely that targeted therapies for a much wider range of cancers are going to emerge within the next decade, but it also seems equally unlikely that chemotherapy, radiation, and surgery will be supplanted anytime soon for most cancers.

So the reader is urged to be cautious when reading of exciting new breakthroughs in genetic medicine. They will certainly come, but there will be disappointments too. Although we all hope for a green light, it is best to pay attention when the amber light is blinking.

References and notes

Chapter 1: Hunting for disease genes

[1] Wikipedia. "Prince Leopold, Duke of Albany." http://en.wikipedia.org/wiki/Prince_Leopold,_Duke_of_Albany. "Death of Prince Leopold." *New York Times*, March 29, 1884. McGonagall, William Topaz. Poemhunter.com. "The Death of Prince Leopold." http://www.poemhunter.com/poem/the-death-of-prince-leopold/#at.

[2] "The Death of Prince Leopold." *Medical Record* 25 (1884): 444.

[3] Wilson, A. N. *The Victorians*. London: Arrow, 2002, p. 25.

[4] Sawaf, Haidi, Adonis Lorenzana, and Lawrence F. Jardine. "Hemophilia A and B." emedicine, http://emedicine.medscape.com/article955590-print.

[5] Gill, Peter et. al. "Identification of the Remains of the Romanov Family by DNA Analysis." *Nature Genetics* 6 (1994): 130–135. Ivanov, Pavel et. al. "Mitochondrial DNA Sequence Heteroplasmy in Grand Duke of Russia Georgij Romanov Establishes the Authenticity of the Remains of Tsar Nicholas II." *Nature Genetics* 12 (1996): 417–420.

[6] Rogaev, Evgeny et. al. "Genomic Identification in the Historical Case of the Nicholas II Royal Family." *Proceedings of the National Academy of Sciences* 106 (2009): 5258–5263.

[7] Rogaev, Evgeny et. al. "Genotype Analysis Identifies the Cause of the Royal Disease." *Science* 326 (2009): 817.

[8] Lanska, Douglas J. "George Huntington (1850–1916) and Hereditary Chorea." *Journal of the History of the Neurosciences* 9 (2000): 76–89.

[9] Huntington, George. "On Chorea." *The Medical and Surgical Reporter* 26 (1872): 317–321.

[10] Davenport, C. B. "Huntington's Chorea in Relation to Heredity and Eugenics." *Proceedings of the National Academy of Sciences of the United* States 1 (1915): 283–285.

[11] Bischoff, Dan. "Of Jersey Ruins and Woody Guthrie." *The Newark Star Ledger*, January 17, 2003.

[12] For an excellent, if older, account of the search for the Huntington's gene, see Chapters 1 and 7 in Bishop, Jerry E., and Michael Waldholz. *Genome: The Story of Our Astonishing Attempt to Map All the Genes in the Human Body*. New York: Simon & Schuster, 1991.

[13] Okun, Michael S., and Nia Thommi. "Americo Negrette (1924–2003): Diagnosing Huntington Disease in Venezuela." *Neurology* 63 (2004): 340–343.

[14] The U.S.-Venezuela Collaborative Research Project and Nancy S. Wexler. "Venezuelan Kindreds Reveal That Genetic and Environmental Factors Modulate Huntington's Disease Age of Onset." *Proceedings of the National Academy of Sciences* 101 (2004): 3498-3503.

[15] Gusella, James F., and Nancy S. Wexler et. al. "A Polymorphic DNA Marker Genetically Linked to Huntington's Disease." *Nature* 306 (1983): 234–238.

[16] The Huntington's Disease Collaborative Research Group. "A Novel Gene Containing a Trinucleotide Repeat That Is Expanded and Unstable on Huntington's Disease Chromosomes." *Cell* 72 (1993): 971–983.

[17] Myriad Genetics. http://www.myriad.com.

[18] Myriad Genetics, Inc. 2009 Annual Report. http://investor.myriad.com/annuals.cfm.

[19] McLellan, Tracy, Lynn B. Jorde, and Mark H. Skolnick. "Genetic Distances Between the Utah Mormons and Related Populations." *American Journal of Human Genetics* 36 (1984): 836–857.

[20] Lewis, Ricki. "Founder Populations Fuel Gene Discovery." *The Scientist* 15 (2001): 8.

[21] Singer, Emily. "Gene Hunting in Canada." *Technology Review* (April 10, 2006). http://www.technologyreview.com/printer_friendly_article.sspx?id=16680&channel=biomedicine§ion=.

[22] Genizon BioSciences. http://ww.genizon.com/.

[23] Siegel, Judy. "Dawning of the Age of Genetic Healing." *The Internet Jerusalem Post*, October 21, 2001. http://cgis.jpost.com/cgi-bin/General/printarticle.cgi?article=/Editions/2001/10/21/Health/Health.36631.html.

[24] Shifman, Sagiv et al. "A Highly Significant Association Between a COMT Haplotype and Schizophrenia." *American Journal of Human Genetics* 71 (2002): 1296–1302.

[25] Fishman, Racheller H. B. "Lack of Funding, Failed Contract Forcing IDgene to Close Doors." *BioWorld International*, March 17, 2004. http://www.accessmylibrary.com/article/print/iGi-114423393.

[26] History of Iceland. http://www.iceland.freewebspace.com/ Timeline: Iceland, BBC News. http://newsvote.bbc.co.uk/mpapps/pagetools/print/news.bbc.co.uk/1/hi/world/europe/countly_profiles/1025288.stm.

[27] Unless otherwise noted, the primary source for much of the discussion in this section concerning Iceland's Health Sector Database and the Health Sector Database Act is the excellent article by David E. Winickoff. "Genome and Nation: Iceland's Health Sector Database and Its Legacy." *Innovations: Technology, Governance, Globalization* 1 (2006): 80–105. There is also an entire book on the deCODE story by Michael Fortun. *Promising Genomics: Iceland and deCODE Genetics in a World of Speculation*. Berkeley, CA: University of California Press, 2008.

[28] "Random Samples." *Science* 279 (1998): 991.

[29] Hlodan, Oksana. "For Sale: Iceland's Genetic History." *Actionbioscience*, June 2000. http://www.actionbioscience.org/genomic/hlodan.html.

[30] GENETarchive. http://www.gene.ch/genet/1999/Mar/msg00032.html.

[31] Jónatnsson, Hróbjartur. "Iceland's Health Sector Database: A Significant Head Start in the Search for the Biological Grail or an Irreversible Error?" *American Journal of Law and Medicine* 26 (2000): 31–67.

[32] Republic of Iceland. "Privacy and Human Rights 2003: Iceland." http://www.privacyinternational.org/survey/phr2003/countries/iceland.htm.

[33] Specter, Michael. "Decoding Iceland." *The New Yorker*, January 18, 1999, pp. 40–51.

[34] Hlodan, Oksana, op. cit., 29.

[35] Lewontin, Richard. "Op-ed/People Are Not Commodities." *The New York Times*, January 23, 1999, A19.

[36] About deCODE genetics. http://www.decode.com/Company/index.php.

[37] deCODEme. http://www.decodeme.com/.

[38] deCODE genetics, Inc. "Files Voluntary Chapter 11 Petition to Facilitate Sale of Assets." http://www.decode.wom/News/news.php?s=32.

[39] Henderson, Mark. "Privacy Fears as DNA Testing Firm deCODE Genetics Goes Bust." *TIMESONLINE*, November 18, 2009. http://www.timesonline.co.uk/tol/news/science/genetics/article6920653.ece?print=yes&randnum=1258549974971.

[40] Ibid.

[41] Ibid.

[42] Wade, Nicholas. "Out of Bankruptcy, Genetics Company Drops Drug Efforts." *The New York Times*, January 21, 2010, page B2.

[43] Kevles, Daniel J. *In the Name of Eugenics*. Cambridge: Harvard University Press, 1985, 1995, pp.232–233.

[44] "Genetic Disease Information—*pronto!*" Human Genome Project Information, last modified July 21, 2008. http://www.ornl.gov/sci/techsources/Human_Genome/medicine/assist.shmtl.

[45] "The Science Behind the Human Genome Project: Basic Genetics, Genome Draft Sequence, and Post-Genome Science." Human Genome Project Information, last modified on March 26, 2008. http://www.ornl.gov/sci/techresources/Human_Genome/project/info.shtml.

[46] Goodier, John L., and Haig H. Kazazian. "Retrotransposons Revisted: The Restraint and Rehabilitation of Parasites." *Cell* 135 (2008): 23–35.

[47] Batzer, Mark A., and Prescott L. Deininger. "Alu Repeats and Human Genomic Diversity." *Nature Reviews Genetics* 3 (2002): 370–379.

Alu1 stands for an enzyme from the bacterium A(rthrobacter) lu(teus) that makes a cut in double-stranded DNA at the sequence AGTC/TCGA. This enzyme is called a restriction enzyme. There are many different bacterial restriction enzymes and they cut at different sequences in DNA, usually four to six base pairs. They are very important tools in molecular biology. Like Alu, each of the first letters in the designation is the first letter of the generic name and the second two are the first two letters of the specific name. Bacteria make these enzymes to defeat incoming virus DNA. They modify the restriction sites in their own DNA so they are not susceptible to attack. Hence, these bacterial defense systems are called restriction-modification systems.

[48] Goodier, John L., and Haig H. Kazazian, op. cit., 46.

[49] Kazazian, Haig H. et al. "Hemophilia A Resulting from de Novo Insertion of L1 Sequences Represents a Novel Mechanism for Mutation in Man." *Nature* 232 (1988): 164–166.

[50] Balancio, Victoria P., Prescott L. Deininger, and Astrid M. Roy-Engel. "Line Dancing in the Human Genome: Transposable Elements and Disease." *Genome Medicine* 1 (2009): 97.1–97.8.

Chapter 2: How genetic diseases arise

[1] Garrod, Archibald E. "The Incidence of Alkaptonuria: A Study in Chemical Individuality." *Lancet* 2 (1902): 1616–1620. Garrod's important contributions to biochemical genetics are reviewed by Alexander G. Bearn and Elizabeth Miller. "Archibald Garrod and the Development of the Concept of Inborn Errors of Metabolism." *Bulletin of the History of Medicine* 53 (1979): 315–328.

[2] Erich von Tschermak, a 26 year-old graduate student in Ghent, Belgium, is often listed as the third person to rediscover Mendel's paper, but Robin Marantz Henig in *The Monk in the Garden* (Boston: Houghton Mifflin, 2001), points out that his part in the rediscovery is usually dismissed today.

[3] See Chapter 21 of Nicholas W. Gillham's *A Life of Sir Francis Galton: From African Exploration to the Birth of Eugenics* (New York: Oxford University Press, 2001) for a detailed discussion of the battle between Mendel's proponents and those of Galton's Ancestral Law.

[4] Ibid., p. 338.

[5] Bearn and Miller, op. cit., 1, pp. 323–24.

[6] Ibid., p. 326–27.

[7] Nobelprize.org. "The Nobel Prize in Physiology or Medicine 1958: George Beadle, Edward Tatum, Joshua Lederberg." http://nobelprize.org/nobel_prizes/medicine/laureates/1958/beadle-lecture.html.

[8] Fernández-Cañón, Jóse M. et al. "The Molecular Basis of Alkaptonuria." *Nature Genetics* 14 (1996): 19–24.

[9] Beltrán-Valero de Bernabé, D. et. al. "Analysis of Alkaptoneuria (AKU) Mutations and Polymorphisms Reveals That the CCC Sequence Motif Is a Mutational Hot Spot in the Homogenisate 1,2 Dioxygenase Gene (*HGO*)." *American Journal of Human Genetics* 64 (1999): 1316–1322.

[10] Stahl, Franklin W. "The Amber Mutants of Phage T4." *Genetics* 141 (1995): 439–442.

[11] Carlson, Bruce. "SNPS—A Shortcut to Personalized Medicine." *GEN: Genetic Engineering & Biotechnology News* 28 (2008), no. 12.

[12] Note that many genetic-testing companies use haplotype to refer to an individual collection of short tandem repeats (STRs) while using haplogroup for SNPs. "Haplotype." http://en.wikipedia.org/wiki/haplotype.

[13] National Human Genome Research Institute. http://www.genome.gov/10001665.

[14] National Institutes of Health. "International HapMap Consortium Publishes Scientific Strategy." *NIH News Advisory*, December 18, 2003. http://www.genome.gov/11509579.

[15] Gibbs, R.A.. "The International HapMap Project." *Nature* 426 (2003): 789–796.

[16] Frazer, K.A. "A Second Generation Human Haplotype Map of Over 3.1 Million SNPs." *Nature* 449 (2007): 851–862.

[17] McCarthy, Mark I. et al. "Genome-Wide Association Studies for Complex Traits: Consensus, Uncertainty and Challenges." *Nature Reviews: Genetics* 9 (2008): 356–369.

[18] Wade, Nicholas. "In the Genome Race, the Sequel Is Personal." *The New York Times*, September 4, 2007, page B1.

[19] Levy S. et al. "The Diploid Genome Sequence of an Individual Human." *PloS Biology* 5 (2007): 2113–2133; Venter, J. Craig. *A Life Decoded*. New York: Viking Press, 2007.

[20] Wade, op. cit., 18.

[21] See Gillham, N.W. op. cit., 3, Chapter 13 for a fuller discussion of pangenesis.

[22] Muller, H. J. "Our Load of Mutations." *American Journal of Human Genetics* 2 (1950): 111–176.

[23] See discussion of genetic load in Kevles, Daniel L. *In the Name of Eugenics*. Cambridge: Harvard University Press, 1995, pp. 259–262.

[24] Kilbey, Brian J. "Charlotte Auerbach (1899–1994)." *Genetics* 141 (1995): 1–5.

[25] Ames, Bruce N., Frank D. Lee, and William E. Durston. "An Improved Bacterial Test System for the Detection and Classification of Mutagens and Carcinogens." *Proceedings of the National Academy of Sciences USA* 70 (1973): 782–786.

[26] Percivall Pott. http://en.wikipedia.org/wiki/Percivall_Pott.

[27] Apple Valley Chimney Sweeps. "About Us and the History of Chimney Sweeps." https://applevalleychimneysweep.com/About_Us.html.

[28] McCann, Joyce, and Bruce N. Ames. "Detection of Carcinogens as Mutagens in the Salmonella/Microsome Test: Assay of 300 Chemicals: Discussion." *Proceedings of the National Academy of Sciences USA* 73 (1976): 950–954.

[29] Florin, Inger et al. "Screening of Tobacco Smoke Constituents for Mutagenicity Using the Ames' Test." *Toxicology* 18 (1980): 219–232.

[30] Ames, Bruce N., H. O, Kammen, and Edith Yamasaki. "Hair Dyes Are Mutagenic: Identification of a Variety of Mutagenic Ingredients." *Proceedings of the National Academy of Sciences of the United States* 72 (1975): 2423–2427. Takkouche, Bahi, Mahyar Etminan, and Agustin Montes-Martinez. "Personal Use of Hair Dyes and Risk of Cancer: A Meta-Analysis." *Journal of the American Medical Association* 293 (2005): 2516–2525. Bolt Hermann M., and Klaus Golka. "The Debate on Carcinogenicity of Permanent Hair Dyes: New Insights." *Current Reviews in Toxicology* 37 (2007): 521–536.

[31] MacGregor, James T., Danial Casciano, and Lutz Müller. "Strategies and Testing Methods for Identifying Mutagenic Risks." *Mutation Research* 455 (2000): 3–20.

[32] UV Index: SunWise Program. http://nsdi.epa.gov/sunwise/uvindex.html.

[33] The Skin Cancer Foundation. "Understanding UVA and UVB." http://www.skincancer.org/Understanding-UVA-and-UVB.html.

[34] Ibid.

[35] Quinton, Paul M. "Physiological Basis of Cystic Fibrosis: A Historical Perspective." *Physiological Reviews* 79, Supplement 1 (1999): S3–S22.

[36] Cystic Fibrosis Research Directions: NIDDK. http://www.wrongdiagnosis.com/artic/cystic_fibrosis_research_directions_niddk.htm.

[37] O'Sullivan, Brian P., and Steven Freedman. "Cystic Fibrosis." *The Lancet* 373 (2009): 1891–1904.

[38] Assael, Baroukh M. et al. "Epiedemiology and Survival Analysis of Cystic Fibrosis in an Area of Intense Screening Over 30 Years." *American Journal of Epidemiology* 66 (2002): 397–401.

[39] Bobadilla, Joseph L. et al. "Cystic Fibrosis: A Worldwide Analysis of CFTR Mutations—Correlation with Incidence Data and Application to Screening." *Human Mutation* 19 (2002): 575–606.

[40] Ferrell, Philip et al. "Discovery of the Principal Cystic Fibrosis Mutation (F508del) in Ancient DNA from Iron Age Europeans." *Nature Precedings*: hdl:10110/npre.2007.1276.1: Posted 29 October 2007 http://precedings.nature.com/users/1e96a0049683c41e24fc029ad9fa9dad.

[41] Quinton, op. cit., 35.

[42] Who Named It? "John Langdon Haydon Down." http://www.whonamedit.com/doctor.cfm/335.html.

[43] Down, J. Langdon H. "Observations on an Ethnic Classification of Idiots." *London Hospital Reports* 3 (1866): 259–262.

[44] Wright, David. *Mental Disability in Victorian England: The Earlswood Asylum 1847–1901*. Oxford: Oxford University Press, 2001, pp.172–174.

[45] For an excellent summary of the work done by Penrose, see Chapter 10 in Kevles, Daniel. Op. cit., 23.

[46] For an excellent summary of the history of human chromosome counting, see Kottler, Malcolm Jay. "From 48 to 46: Cytological Technique, Preconception, and the Counting of Human Chromosomes." *Bulletin of the History of Medicine* 48 (1974): 465–502.

[47] Answers.com. "Chromosomal Banding." http://www.answers.com/topic/chromosomal-banding.

[48] Wikipedia. "Jérôme Lejeune." http://en.wikipedia.org/wiki/ Jérôme_Lejeune.

[49] "What Is Down syndrome?" http://www.downsyn.com/whatisds.php.

[50] Mansfield, Caroline et al. "Termination Rates After Prenatal Diagnosis of Down Syndrome, Spina Bifida, Anencephaly, and Turner and Klinefelter Syndromes: A Systematic Literature Review." *Prenatal Diagnosis* 19 (1999): 808–812.

[51] Will, George F. "Eugenics By Abortion." *Washington Post*, April 14, 2005, page A27.

[52] Cuniff, Christopher et al. "Health Supervision for Children With Down Syndrome." *Pediatrics* 107 (2001): 442–449.

[53] Ibid.

[54] Quenby, Siobhan et al. "Recurrent Miscarriage: A Defect in Nature's Quality Control?" *Human Reproduction* 17 (2002): 1959–1963.

[55] Nicolaidis, Peter, and Michael B. Petersen. "Origin and Mechanisms of Non-Disjunction in Human Autosomal Trisomies." *Human Reproduction* 13 (1998): 313–319.

[56] Morton, N. E. et al. "Maternal Age in Trisomy." *Annals of Human Genetics* 52 (1988): 227–236.

[57] Lyon, Mary. "Gene Action in the X-chromosome of the Mouse (Mus musculus L.)." *Nature* 190 (1961): 372–373.

[58] Kevles, op. cit., 23, pp. 242–245.

[59] Lanfranco, Fabio et al. "Klinefelter's Syndrome." *The Lancet* 364 (2004): 273–283.

[60] "Male Tortoiseshell Cat 'Genetically Impossible." *Daily Telegraph*, August 26, 2009.

[61] Wutz, Anton. "Xist Function: Bridging Chromatin and Stem Cells." *Trends in Genetics* 23 (2007): 457–464. Wutz, Anton, and Joost Gribnau. "X Inactivation Xplained." *Current Opinion in Genetics and Development* 17 (2007): 387–393.

[62] Angelman Syndrome Foundation, Inc. "Harry Angelman and the History of AS." http://www.angelman.org/stay-informed/facts-about-angelman-syndrome---7th-edition/harry-angelman-and-the-history-of-as/.

[63] Ibid.

[64] Genetics Home Reference: Your Guide to Understanding Genetic Conditions. "Angelman syndrome." http://ghr.nlm.nih.gov/condition=angelmansyndrome.

[65] Genetics Home Reference: Your Guide to Understanding Genetic Conditions. "Prader-Willi syndrome." http://ghr.nlm.nih.gov/condition=praderwillisyndrome.

[66] Horsthemke, Bernhard, and Joseph Wagstaff. "Mechanisms of Imprinting of the Prader-Willi/Angelman Regon." *American Journal of Medical Genetics* Part A 146A (2008): 2041–2052. Note that the paternal chromosome 15 region deleted in Prader-Willi syndrome also contains around 70 genes encoding C/D box snoRNAs. These small RNAs guide methylation of other RNAs including ribosomal RNA and transfer RNAs.

[67] Genetics Home Reference. "OCA2." http://ghr.nlm.nih.gov/gene=oca2.

[68] Cloud, John. "Why Your DNA Isn't Your Destiny." *Time*, January 6, 2010.

Chapter 3: Ethnicity and genetic disease

[1] Wikipedia. "J. B. S. Haldane." http://en.wikipedia.org/wiki/J._B._S._Haldane. "The Unofficial Stephen Jay Gould Archive: People: J.B.S. Haldane." http://www.stephenjaygould.org/people/john_haldane.html.

[2] Lederberg, Joshua. "J. B. S. Haldane (1949) on Infectious Disease and Evolution." *Genetics* 153 (1999): 1–3. Also see Haldane, J. B. S. "The Rate of Mutation of Human Genes." *Hereditas* 35 (1949): 267–273.

[3] "A Brief History of Sickle Cell Disease." http://sickle.bwh.harvard.edu/scd_history.html. Innvista. "Sickle Cell History." http://www.innvista.com/health/ailments/anemias/sickhist.htm. The Africa Guide. "Ghana." http://www.africaguide.com/country/ghana/. Ethnologue, Languages of the World "Languages of Ghana." http://www.ethnologue.com/show_country.asp?name=ghana.

[4] GhanaWeb. "Early European Contact and the Slave Trade." http://www.ghanaweb.com/GhanaHomePage/history/slave-trade.php.

[5] Washington University Biology Department. "Biogeography and Ecology of Sickle Cell Anemia." http://www.nslc.wustl.edu/sicklecell/part3/biogeography.html.

[6] Allison, A. C. "Protection Afforded by Sickle-Cell Trait Against Subtertian Malarial Infection." *British Medical Journal* 1 (1954): 290–294.

[7] Lederberg, op. cit., 2.

[8] Encyclopedia Brittanica. "James Bryan Herrick." http://www.britannica.com/EBchecked/topic/263733/James-Bryan-Herrick. Wikipedia. "James B. Herrick." http://en.wikipedia.org/wiki/James_B._Herrick.

[9] Williams, T. N. et al. "An Immune Basis for Malaria Protection by the Sickle Cell Trait." *PloS Medicine* 2 (2005): 0442–0445.

[10] Greenwood, Brian M. et al. "Malaria." *The Lancet* 365 (2005): 1487–1498.

[11] There is a nice account of the unraveling of the molecular basis of sickle cell anemia in Chapter 15 of Daniel Kevles's book *In the Name of Eugenics*. Cambridge: Harvard University Press, 1995.

[12] There are all sorts of sites dealing with sickle cell disease and its symptoms on the Internet. A good succinct description of this genetic disease and many others is to be found in Genetics Home Reference. http://ghr.nlm.nih.gov/. For a far more detailed description, see Bender, M. A., and W. Hobbs. National Center for Biotechnology Information, U.S. National Library of Medicine. Gene Reviews. "Sickle Cell Disease." http://www.ncbi.nlm.nih.gov/bookshelf/br.fcgi?book=gene&part=sickle.

[13] Platt, O. S. et al. "Mortality in Sickle Cell Disease—Life Expectancy and Risk Factors for Early Death." *New England Journal of Medicine* 330 (1994): 1639–1644. Quinn, Charles T., Zora R. Rogers, and George R. Buchanan. "Survival of Children with Sickle Cell Disease." *Blood* 103 (2004): 4023–4027.

[14] Bender and Hobbs, op. cit., 12.

[15] Strouse, John J. et al. "Hyroxyurea for Sickle Cell Disease: A Systematic Review for Efficacy and Toxicity in Children." *Pediatrics* 122 (2008): 1332–1342.

[16] Platt, Orah S. "Hydroxyurea for the Treatment of Sickle Cell Anemia." *The New England Journal of Medicine* 358 (2008): 1362–1369.

[17] Ibid.

[18] Jones, Stephen R., Richard A. Binder, and Everett M. Donowho, Jr. "Sudden Death in Sickle Cell Trait." *New England Journal of Medicine* 282 (1970): 323–325.

[19] Kevles, op. cit., 11, pp. 278, 300.

[20] Ibid. 256, 278.

[21] Weiss, Robin Elise. About.com: Pregnancy & Childbirth. "Newborn Screening Recommendations by State." http://pregnancy.about.com/od/newbornbabies/a/newbornstate.htm.

[22] National Heart, Lung, and Blood Institute, National Institutes of Health. "Sickle Cell Research for Treatment and Cure." http://www.nhlbi.nih.gov/resources/docs/scd30/scd30.pdf.

[23] Dodd, Dennis. CBS Sports.com. "Fatalities Continue—and So Does Ignorance of Sickle Cell Trait." http://www.cbssports.com/collegefootball/story/11742560.

[24] National Athletic Trainers' Association. 2007. "Consensus Statement: Sickle Cell Trait and the Athlete." http://www.nata.org/statements/consensus/sicklecell.pdf.

[25] Bouchette, Ed. "On the Steelers: Holmes' Sickle Cell Not Affected in Denver." *Pittsburgh Post-Gazette*, November 3, 2009.

[26] NFL Fanhouse. "No Ryan Clark, but Same Steel Curtain." http://nfl.fanhouse.com/2009/11/10/no-ryan-clark-but-same-steel-curtain/.

[27] Bouchette, op. cit., 25.

[28] Alabdulaali, M. K. "Sickle Cell Disease Patients in Eastern Province of Saudi Arabia Suffer Less Severe Acute Chest Syndrome Than Patients with African Haplotypes." *Annals of Thoracic Medicine* 2 (2007): 158–162.

[29] Wikipedia. "History of Malaria." http://en.wikipedia.org/wiki/History_of_malaria.

[30] Who Named It? "Thomas Benton Cooley." http://www.whonamedit.com/doctor.cfm/1931.html Who Named It? "Cooley's Anaemia." http://www.whonamedit.com/doctor.cfm/1931.html.

[31] Nobelprize.org. "George H. Whipple: The Nobel Prize in Physiology or Medicine 1934." http://nobelprize.org/nobel_prizes/medicine/laureates/1934/whipple-bio.html.

[32] Hill, A. V. S. et al. "β Thalassemia in Melanesia: Association With Malaria and Characterization of a Common Variant (IVS-1 at G->C)." *Blood* 72 (1988): 9–14. Genetics Home Reference. "Beta Thalassemia." http://ghr.nlm.nih.gov/condition=betathalassemia

[33] Weatherall, David et al. "Inherited Disorders of Hemoglobin." Chap. 34 in *Disease Control Priorities in Developing Countries*. 2nd ed. Oxford University Press and the World Bank, 2006. http://files.dcp2.org/pdf/DCP/DCP34.pdf.

[34] Tebbutt, Tom. "Sampras Battles Inherited Anemia." *Toronto Globe and Mail*, September 24, 1996.

[35] Sampras, Pete with Peter Bodo. *A Champion's Mind: Lessons from a Life in Tennis*. New York: Random House, 2008, p. 181.

[36] Neufeld, Ellis J. "Oral Chelators Deferasirox and Deferiprone for Transfusional Iron Overload in Thalassemia Major: New Data, New Questions." *Blood* 107 (2006): 3436–3441.

[37] Weatherall, David, op. cit., 33, Table 34.1.

[38] Guterman, Lila, and Francis X. Rocca. "Choosing Eugenics: How Far Will Nations Go to Eliminate Genetic Disease?" *The Chronicle of Higher Education* 49 (2003): A22–24.

[39] Mockenhaupt, Frank P. et al. "α^+-thalassemia Protects from Severe Malaria in African Children." *Blood* 104 (2004): 2003–2006. Williams, Thomas N. et al. "Both Heterozygous and Homozygous α^+ thalassemias Protect Against Severe and Fatal *Plasmodium falciparum* Malaria on the Coast of Kenya." *Blood* 106 (2005): 368–371.

[40] Fowkes, Freya J. I. et al. "Increased Microerythrocyte Count in Homozygous α^+-Thalassemia Contributes to Protection Against Severe Malarial Anaemia." *PloS Medicine* 5 (2008): 0494–0501.

[41] Fairhurst, R. M. et al. "Abnormal Display of PfEMP-1 on Erythrocytes Carrying Haemoglobin C May Protect Against Malaria." *Nature* 435 (2005): 1117–1121.

[42] Chotivanich, K. et al. "Hemoglobin E: A Balanced Polymorphism Protective Against High Parasitemias and thus Severe *P. falciparum* Malaria." *Blood* 100 (2002): 1172–1176.

[43] Kalow, Werner. "Pharmacogenetics: Its Place in Medicine and Biology." *Journal of Pharmacy Practice* 5 (1992): 312–316.

[44] Genetics Home Reference. "Glucose-6-Phosphate Dehydrogenase Deficiency." http://ghr.nlm.nih.gov/condition=glucose6phosphatedehydrogenasedeficiency/show/print Mason, Philip J., José M. Bautista, and Florinda Gilsanz. "G6PD Deficiency: The Genotype-Phenotype Association." *Blood Reviews* 21 (2007): 267–283.

[45] WHO Working Group. "Glucose-6-Phosphate Dehydrogenase Deficiency." *Bulletin of the World Health Organization* 67 (1989): 801–811.

[46] Tishkoff, Sarah A. et al. "Haplotype Diversity and Linkage Disequilibrium at Human *G6PD*: Recent Origin of Alleles That Confer Malarial Resistance." *Science* 293 (2001): 455–462.

[47] Belsey, Mark A. "The Epidemiology of Favism." *Bulletin of the World Health Organization* 48 (1973): 1–13.

[48] Cornell University Department of Animal Science. "Vicine and Covicine: Plants Poisonous to Livestock." http://www.ansci.cornell.edu/plants/toxicagents/vicine.html.

[49] Snow, Robert W. et al. "The Global Distribution of Clinical Episodes of *Plasmodium falciparum* Malaria." *Nature* 434 (2005): 214–217.

[50] Wikipedia. "Francisco J. Ayala." http://en.wikipedia.org/wiki/Francisco_J._Ayala.

[51] Rich, Stephen M. et al. "The Origin of Malignant Malaria." *Proceedings of the National Academy of Sciences*. 106 (2009): 14902–14907. The erythrocyte-binding antigen on the surface of the plasmodium merozoite is called EBA175. EBA175 recognizes sialic acids (Sias) presented on clusters of O-linked glycans attached to erythrocyte surface glycophorin A. Thus, the Sia-capped glycoprotein GYPA is the primary binding target for EBA175. Mammalian erythrocytes commonly have two Sias N-glycolylneuraminic acid (Neu5Gc) and N-acetylneuraminic acid (Neu5Ac). Neu5Gc is synthesized from Neu5Ac. *Plasmodium reichenowi* EBA175 binds preferentially to Neu5Gc, but a mutation in the human *CMAH* gene blocks synthesis of Neu5Gc from Neu5Ac protecting humans against *P. reichenowi*. However, *P. falciparum* EBA175 does bind to Neu5Ac and so its merozoites easily infect human erythrocytes.

[52] Galvani, Alison P., and Montgomery Slatkin. "Evaluating Plague and Smallpox as Historical Selective Pressures for the CCR5-Δ32 HIV-Resistance Allele." *Proceedings of the National Academy of Sciences* 100 (2003): 15276–15279.

[53] Halliday, Stephen. *The Great Stink of London: Sir Joseph Bazalgette and the Cleansing of the Victorian Metropolis*. London: Sutton Publishing Ltd., 1999, p. ix.

[54] Wilson, A. N. *The Victorians*. London: Arrow Books, 2003, p. 155. Note that typhoid and waterborne typhus are the same disease.

[55] Ibid. p. 243.

[56] Halliday, op. cit., 53, p. xii–xiii.

[57] Gabriel, Sherif E. et al. "Cystic Fibrosis Heterozygote Resistance to Cholera Toxin in the Cystic Fibrosis Mouse Model." *Science* 266 (1994): 107–109.

[58] Cuthbert, A. W. et al. "The Genetic Advantage Hypothesis in Cystic Fibrosis Heterozygotes: A Murine Study." *Journal of Physiology* 482.2 (1995): 449–464.

[59] Högenauer, Christoph. "Active Intestinal Chloride Secretion in Human Carriers of Cystic Fibrosis Mutations: An Evaluation of the Hypothesis That Heterozygotes Have Subnormal Active Intestinal Chloride Secretion." *American Journal of Human Genetics* 67 (2000): 1422–1427.

[60] Pier, Gerald B. et al. "Salmonella Typhi Uses CFTR to Enter Intestinal Epithelial Cells." *Nature* 393 (1998): 79–82.

[61] Tsui, I. S. M. et al. "The Type IVB Pili of Salmonella Enterica Serovar Typhi Bind to the Cystic Fibrosis Transmembrane Conductance Regulator." *Infectious Immunology* 71 (2003): 6049–6050.

[62] Poolman, Eric M., and Alison P. Galvani. "Evaluating Candidate Agents of Selective Pressure for Cystic Fibrosis." *Journal of the Royal Society Interface* 4 (2007): 91–98.

[63] Tobacman, Joanne K. "Does Deficiency of Arylsulfatase B Have a Role in Cystic Fibrosis?" *Chest* 123 (2003): 2130–2139.

[64] Provine, William B. "Ernst Mayr: Genetics and Speciation." *Genetics* 167 (2004): 1041–46.

[65] Scriver, Charles R. "Human Genetics: Lessons from Quebec Populations." *Annual Review of Genomics and Human Genetics* 2 (2001): 69–101. Unless otherwise noted by another citation, the discussion of hereditary diseases in Quebec is derived from this reference.

[66] Genetics Home Reference. "Hypercholesterolemia." http://ghr.nlm.nih.gov/condition =hypercholesterolemia.

[67] Genetics Home Reference. "Tay-Sachs Disease." http://ghr.nlm.nih.gov/condition= taysachsdisease.

[68] Scriver, op. cit., 65.

[69] Lacassie, Yves, and Luisa Flórez. LSU Health Center. Genetics and Louisiana Families. "Some Genetic Disorders Among Acadian People." http://www.med-school.lsuhsc.edu/genetics_center/louisiana/article_geneticdisorders.htm.

[70] McDowell, Geraldine A. et al. "The Presence of Two Different Infantile Tay-Sachs Disease Mutations in a Cajun Population." *Journal of Human Genetics* 51 (1992): 1071–1077.

[71] Center for Jewish Genetic Diseases, Mount Sinai Hospital. http://www.mssm.edu/ research/programs/jewish-genetics-disease-center.

[72] Chicago Center for Jewish Genetic Disorders. http://www.jewishgenetics.org.

[73] Jewish Genetic Diseases Center of Greater Phoenix. http://www.jewishgeneticsphx. org.

[74] Klein, Aaron. "Genetic Screening Causes Controversy." *Yeshiva University Commentator* 62, issue 10. http://commie.droryikra.com/archives/v62iA/news/ doryesharim.html. "Dor Yeshorim." *The Jewish Press*, November 25, 2009. http: //www.jewishpress.com/pageroute.do/41623.

[75] The Jewish Federations of North America. Jewish Life. "NJPS 2000-01 Report." http://www.jewishfederations.org/page.aspx?id=45894.

[76] Shimo-Barry, Alex. "Modern Matchmaker in Race and Ethnicity in the New Urban America." http://web.jrn.columbia.edu/studentwork/race/2002/gene-shimo.shtml.

[77] Rosen, Christine. "Eugenics—Sacred and Profane." *The New Atlantis*, Summer 2003, 79–89.

[78] Kaplan, Feige. "Tay-Sachs Disease Carrier Screening: A Model for Prevention of Genetic Disease." *Genetic Testing* 2 (1998): 271–292.

[79] Oster, Harry. "A Genetic Profile of Contemporary Jewish Populations." *Nature Reviews Genetics* 2 (2001): 891–898.

[80] Rotter, Jerome J., and Jared M. Diamond. "What Maintains the Frequencies of Human Genetic Diseases?" *Nature* 329 (1987): 289–290.

[81] Spyropoulos B. et al. "Heterozygous Advantage in Tay-Sachs Carriers? *American Journal of Human Genetics* 33 (1981): 375–380.

[82] Diamond, Jared. "Jewish Lysosomes." *Nature* 368 (1994): 291–292.

[83] Cochran, Gregory, Jason Hardy, and Henry Harpending. "Natural History of Ashkenazi Intelligence." *Journal of Biosocial Science* 38 (2006): 659–693.

[84] Wade, Nicholas. "Researchers Say intelligence and Diseases May Be Linked in Ashkenazic Genes." *The New York Times*, June 3, 2005, A21.

[85] Risch, Neil et al. "Geographic Distribution of Disease Mutations in the Ashkenazi Jewish Population Supports Genetic Drift over Selection." *American Journal of Human Genetics* 72 (2003): 812–822.

[86] Senior, Jennifer. "Are Jews Smarter?" *New York Magazine*. October 16, 2005. http://nymag.com/nymetro/news/culture/features/1478/.

[87] Slatkin, Montgomery. "A Population-Genetic Test of Founder Effects and Implications for Ashkenazi Jewish Diseases." *American Journal of Human Genetics*. 75 (2004): 282–293.

[88] Popper, Nathaniel. "Study's Claim On Intelligence of Ashkenazim Spurs a Debate." *Jewish Daily Forward*, June 10, 2005. http://www.forward.com/articles/3623/.

Chapter 4: Susceptibility genes and risk factors

[1] Angier, Natalie. "Scientists Detect a Genetic Key to Alzheimer's." *The New York Times*, August 13, 1993. For a review of the early findings, see Strittmatter, Warren J., and Allen D. Roses. "Apolipoprotein E and Alzheimer's Disease." *Annual Review of Genetics* 19 (1996): 53–77. Wikipedia. "Apolipoprotein E." http://en.wikipedia.org/wiki/Apolipoprotein_E.

[2] *Web*MD. Mojica, Ray Camille "Tolerate Lactose Intolerance." www.webmd.com/diet/features/tolerate-lactose-intolerance.

[3] Swallow, Dallas M. "Genetics of Lactase Persistence and Lactose Intolerance." *Annual Review of Genetics* 37 (2003): 197–219.

[4] These mutations do not arise in the gene encoding lactase (*LCT*), but are upstream in introns of a gene called *MCM6*. The variant that prevails in Caucasian populations is called C/T(-13910) located in intron 13 of the *MCM6* gene. This base pair change converts the sequence into an enhancer that binds the OCT1 transcription factor. This probably stimulates lactase mRNA transcription. OMIM. "MIM ID *601806 Minichromosome Maintenance, S. pombe, Homolog of, 6: MCM6." http://www.ncbi.nlm.nih.gov/entrez/dispomim.cgi?id=601806&a=601806_AllelicVariant0001.

[5] Itan, Yuval et al. "The Origins of Lactase Persistence in Europe." *PloS Computational Biology* 5 (2009) 8: e1000491. doi:10.1371.pcbi.1000491. http://www.ploscompbiol.org/article/info:doi%2F10.1371%2Fjournal.pcbi.1000491.

[6] Tishkoff, Sarah A. et al. "Convergent Adaptation of Lactase Persistence in Africa and Europe." *Nature Genetics* 39 (2007): 31–40. Note that other mutations in introns of the MCM6 enhance lactase production in certain African pastoralist populations.

[7] Wikipedia. "Nicotine." http://en.wikipedia.org/wiki/Nicotine.

[8] Coon, Minor J. "Cytochrome P450: Nature's Most Versatile Biological Catalyst." *Annual Review of Pharmacology and Toxicology* 45 (2005): 1–25. Rautio, A. "Polymorphic CYP2A6 and Its Clinical and Toxicological Significance." *The Pharmacogenomics Journal* 3 (2003): 6–7.

[9] Munafò, Marcus, and Elaine C. Johnstone. "Genes and Cigarette Smoking." *Addiction* 103 (2008): 893–904.

[10] These papers identify rs8034191 and rs1051730 as increasing the risk of lung cancer. Amos, Christopher I. et al. "Genome-Wide Association Scan of Tag SNPs Identifies a Susceptibility Locus for Lung Cancer at 15q25.1." *Nature Genetics* 40 (2008): 616–622. Hung, Rayjean J. et al. "A Susceptibility Locus for Lung Cancer Maps to Nicotinic Acetylcholine Receptor Subunit Genes on 15q25." *Nature* 452 (2008): 633–637.

[11] The following papers all document the relationship of rs16969968 to nicotine addiction. Saccone, Scott et al. "Cholinergic Nicotinic Receptor Genes Implicated in a Nicotine Dependence Association Study Targeting 348 Candidate Genes." *Human Molecular Genetics* 16 (2007): 36–49. Weiss, Robert et al. "A Candidate Gene Approach Identifies the *CHRNA5-A3-B4* Region as a Risk Factor for Age-Dependent Nicotine Addiction." *PloS Genetics* 4 (2008): e1000125. doi:10.1371/journal.pgen.1000125. Saccone, Nancy L. et al. "The *CHRNA5-CHRNA3-CHRNB4* Nicotinic Receptor Subunit Gene Cluster Affects Risk for Nicotine Dependence in African Americans and in European Americans." *Cancer Research* 69 (2009): 6848–6856.

[12] The following paper documents the relationship rs1051730 to nicotine addiction. Thorgeirsson, Thorgeir E. et al. "A Variant Associated with Nicotine Dependence, Lung Cancer, and Peripheral Arterial Disease." *Nature* 452 (2008): 638–641.

[13] Saccone, Nancy L. et al. "The *CHRNA5-CHRNA3-CHRNB4* Nicotinic Receptor Subunit Gene Cluster Affects Risk for Nicotine Dependence in African Americans and in European Americans." *Cancer Research* 69 (2009): 6848–6856.

[14] Galvan, Antonella, and Tommaso A. Dragani. "Nicotine Dependence May Link the 15q25 Locus to Lung Cancer Risk." *Carcinogenesis Advance Access*, November 12, 2009. http://carcin.oxfordjournals.org/cgi/reprint/bgp282v1.pdf.

[15] O'Laughlin, J. et al. "Genetically Decreased *CYP2A6* and the Risk of Tobacco Dependence: A Prospective Study of Novice Smokers." *Tobacco Control* 13 (2004): 422–428.

[16] Ariyoshi, Noritaka et al. "Genetic Polymorphism of CKYP2A6 Gene and Tobacco-Induced Lung Cancer Risk in Male Smokers." *Cancer Epidemiology, Biomarkers & Prevention* 11 (2002): 890–894.

[17] Munafò and Johnstone, op. cit., 9.

[18] Ibid.

[19] Gelernter, Joel et al. "Haplotype Spanning *TTC12* and *ANKK1*, Flanked By the *DRD2* and *NCAM1* Loci, Is Strongly Associated to Nicotine Dependence in Two Distinct American Populations." *Human Molecular Genetics* 15 (2006): 3498–3507.

[20] Newsom, Ainsley. "Tracking the Smoking Gene." *The Times*, November 19, 2005.

[21] Ibid.

[22] Ibid.

[23] Ibid.

[24] g-Nostics. http://www.g-nostics.com/.

[25] Ibid.

[26] Profita, Hillary. "And In The End, Peanut Butter Was Not To Blame." Public Eye CBS News, May 12, 2006. http://www.cbsnews.com/sections/publiceye/main500486.shtml?keyword=kiss+of+death.

[27] Bhat, Devika. "The Kiss of Death for Girl with Nut Allergy." *The Times*, November 29, 2005. http://www.timesonline.co.uk/tol/news/world/article597704.ece.

[28] "'Fatal Kiss' Puts Spotlight on Food Allergies." ABC News, November 30, 2005. http://abcnews.go.com/WNT/Health/story?id=1360248Parke.

[29] "Quebec Teen Died of Asthma, Not Nutty Kiss." CBC News, May 11, 2006. http://www.cbc.ca/canada/story/2006/05/11/asthma-death.html.

[30] National Heart Lung and Blood Institute. "Diseases and Conditions Index: Asthma." http://www.nhlbi.nih.gov/health/dci/Diseases/Asthma/Asthma_WhatIs.html.

[31] Dust Mites, Asthma and Allergy Foundation of America. http://www.aafa.org/print.cfm?id=9&sub=22&cont=315#

[32] Bach, Jean-Francois. "The Effect of Infections on Susceptibility to Autoimmune and Allergic Diseases." *The New England Journal of Medicine* 347 (2002): 911–920.

[33] Beasley, Richard. "The Burden of Asthma with Specific Reference to the United States." *Journal of Allergy and Clinical Immunology* 109 (2002): S482–S489.

[34] Ridley, Matt. *Genome: The Autobiography of a Species in 23 Chapters*. New York: Harper Collins, 1999, p. 67.

[35] For a simple summary, see Genentech. "IgE and Its Role in Asthma." http://www.gene.com/gene/products/education/immunology/ige.html.

[36] The discussion that follows is based largely on the following review. Bossé, Yohan, and Thomas J. Hudson. "Toward a Comprehensive Set of Asthma Susceptibility Genes." *Annual Review of Medicine* 58 (2007): 171–184.

[37] McGrath, John A., and Jouni Uitto. "The Filaggrin Story: Novel Insights into Skin-Barrier Function and Disease." *Trends in Molecular Medicine* 14 (2007): 20–27.

[38] Bossé and Hudson, op. cit., 36 for a discussion.

[39] Farfel, Zvi, Henry R. Bourne, and Taroh Liri. "The Expanding Spectrum of G Protein Diseases." *The New England Journal of Medicine* 340 (1999): 1012–1020.

[40] Ose, Leiv. "The Real Code of Leonardo da Vinci." *Current Cardiology Reviews* 4 (2008): 60–62.

[41] Wikipedia. "Familial Hypercholesterolemia." http://en.wikipedia.org/wiki/Familial_hypercholesterolemia.

[42] There is a nice account of their discovery in Bishop, Jerry E., and Michael Waldholz. *Genome*. New York: Simon and Schuster, 1990, Chapter 9.

[43] Genetics Home Reference. "Hypercholesterolemia." http://ghr.nlm.nih.gov/condition=hypercholesterolemia.

[44] Genetics Home Reference. "APOB." http://ghr.nlm.nih.gov/gene=apob/show/print.

[45] Genetics Home Reference, op. cit., 43.

[46] Medline Plus. "Coronary Artery Disease." http://www.nlm.nih.gov/medlineplus/coronaryarterydisease.html.

[47] Hamsten, A., and P. Eriksson. "Identifying the Susceptibility Genes for Coronary Artery Disease: From Hyperbole Through Doubt to Cautious Optimism. *Journal of Internal Medicine* 263 (2008): 538–552.

[48] Wikipedia. "Diabetes Mellitus." http://en.wikipedia.org/wiki/Diabetes_mellitus.

[49] Grant, Struan F. A., and H. Hakonarson. "Genome-Wide Association Studies in Type 1 Diabetes." *Current Diabetes Reports* 9 (2009): 157–163.

[50] Roep, B. O. "Missing Links." *Nature* 450 (2007): 799–800. Nejentsev, Sergey et al. "Localization of Type 1 Diabetes Susceptibility to the MHC Class 1 Genes HLA-B and HLA-A." *Nature* 450 (2007): 887–892.

[51] Grant and Hakonarson, op. cit., 49.

[52] National Institutes of Health. Fact Sheet. "Type 2 Diabetes." http://www.nih.gov/about/researchresultsforthepublic/Type2Diabetes.pdf.

[53] Muolo, Deborah M., and Christopher B. Newgard. "Molecular and Metabolic Mechanisms of Insulin Resistance and β-cell Failure in Type 2 Diabetes." *Nature Reviews: Molecular Cell Biology* 9 (2008): 193–205.

[54] Staiger, Harald et al. "Pathomechanisms of Type 2 Diabetes Genes." *Endocrine Reviews* 30 (2009): 557–585.

[55] Op. cit., 52.

[56] National Institute of Diabetes and Digestive and Kidney Diseases, National Institutes of Health. "Monogenic Forms of Diabetes: Neonatal Diabetes Mellitus and Maturity-Onset Diabetes of the Young." http://www.diabetes.niddk.nih.gov/dm/pubs/mody/#9.

[57] Reich, David E., and Eric S. Lander. "On the Allelic Spectrum of Human Disease." *Trends in Genetics* 17 (2001): 502–510.

[58] National Institute on Aging. National Institutes of Health. "Alzheimer's Disease Genetics Fact Sheet." http://www.nia.nih.gov/NR/rdonlyres/3C4B634E-A2D8-4415-927F-4B79BEC47EA6/11207/84206ADEARFactsheetGeneticsFINAL08DEC23.pdf

[59] Bodmer, Walter, and Carolina Bonilla. "Common and Rare Variants in Multifactorial Susceptibility to Common Diseases." *Nature Genetics* 40 (2008): 695–701.

[60] Wikipedia. "Progeria." http://en.wikipedia.org/wiki/Progeria.

[61] Wade, Nicholas. "A New Way to Look for Disease' Genetic Roots." *New York Times*, January 26, 2010.

[62] Dickson, Samuel P. et al. "Rare Variants Create Synthetic Genome-Wide Associations." *PloS Biology* 8 (2010): 1–12.

[63] Ridley, Matt. "The Failed Promise of Genomics." *The Wall Street Journal*. October 9–10, 2010, C4.

[64] Wade, op. cit., 61.

[65] International Classification of Diseases (ICD) WHO: History. "History of the Development of the ICD." http://www.who.int/classifications/icd/en/HistoryOfICD.pdf.

[66] Goh, Kwang-Il et al. "The Human Disease Network." *Proceedings of the National Academy of Sciences of the United States* 104 (2007): 8685–8690.

Chapter 5: Genes and cancer

[1] National Cancer Institute. "Milestone (1971) National Cancer Act of 1971." http://dtp.nci.nih.gov/timeline/noflash/milestones/M4_Nixon.htm.

[2] Rockoff Alan, and William C. Shiel Jr. MedicineNet.com. "Actinic Keratosis (Solar Keratosis)." http://www.medicinenet.com/script/main/art.asp?articlekey=18893&pf=3&page=1. Spencer, James M., and Amy Lynn Basile. Medscape: CDC Commentary Series. "Actinic Keratosis." http://emedicine.medscape.com/article/1099775-overview.

[3] National Cancer Institute. "Head and Neck Cancer: Questions and Answers." http://www.cancer.gov/cancertopics/factsheet/Sites-Types/head-and-neck/print?page=&keyword=.

[4] Weinberg, Robert A. *The Biology of Cancer*. New York: Garland Science, Taylor & Francis Group, LLC, 2007, pp. 28–33.

[5] Lacour, J. P. "Carcinogenesis of Basal Cell Carcinomas: Genetics and Molecular Mechanisms." *British Journal of Dermatology* 146 (2002), Suppl. 61: 17–19.

[6] Ibid. Also see Weinberg, op. cit., 4, p.779. A protein called hedgehog binds to the patched receptor. In the absence of hedgehog, patched inhibits another membrane protein called smoothened. This prevents smoothened from protecting a protein called gli from cleavage in the cytoplasm. One of the cleavage products of gli moves to the nucleus, where it acts as a transcriptional repressor. When hedgehog binds to patched, smoothened is not inhibited and protects gli from cleavage. The intact gli protein then moves to the nucleus, where it stimulates transcription and expression of certain genes resulting in stimulation of cell division. Thus, acting through patched, hedgehog can modulate this process. Patched mutants cannot prevent smoothened from protecting gli, meaning that the gli protein continues to stimulate transcription and cell growth when it should not. Similarly, certain smoothened mutants fail to respond to patched, so the gli protein is protected and growth stimulation continues when it is supposed to have stopped.

[7] Sherr, Charles J. "Principles of Tumor Suppression." *Cell* 116 (2004): 235–246.

[8] Genetics Home Reference. "Li-Fraumeni Syndrome." http://ghr.nlm.nih.gov/condition=lifraumenisyndrome.

[9] Weinberg, op. cit., 4, p. 34.

[10] Genetics Home Reference. "Gorlin Syndrome." http://ghr.nlm.nih.gov/condition=gorlinsyndrome.

[11] Bishop, Jerry E., and Michael Waldholz. *Genome*. New York: Touchstone, Simon & Schuster, 1990, pp. 135–145.

[12] Toll, A. et al. "MYC Numerical Aberrations in Actinic Keratosis and Cutaneous Squamous Cell Carcinoma." *British Journal of Dermatology* 161 (2009): 1112–1118. *MYC* encodes a protein known as a transcription factor. When the MYC protein forms a complex with another protein called MAX, the MYC-MAX protein complex promotes transcription and cell division. As cell differentiation proceeds, a new complex is formed between a protein called MAD and MAX that inhibits cell division. Hence, when the *MYC* gene is either deregulated or amplified, it behaves as an oncogene because excess MYC accumulates and encourages cell division by continuing to complex with MAX when it is not supposed to.

[13] Weinberg, op. cit., 4, pp. 284–288.

[14] Australasian College of Dermatology. "A–Z of Skin: Moles & Melanoma." http://www.dermcoll.asn.au/public/a-z_of_skin-moles_melanoma.asp.

[15] Cancer. Net. American Society of Clinical Oncology. "The Genetics of Melanoma." http://www.cancer.net/patient/All+About+Cancer/Genetics/The+Genetics+of+Melanoma.

[16] Chudnovsky, Yakov et al. "Melanoma Genetics and the Development of Rational Therapeutics." *Journal of Clinical Investigation* 115 (2005): 813–824.

[17] National Cancer Institute. Surveillance Epidemiology and End Results. "SEER Stat Fact Sheets: Melanoma of the Skin." http://seer.cancer.gov/statfacts/html/melan.html.

[18] Jablonski, Nina G., and George Chaplin. "The Evolution of Human Skin Color." *Journal of Human Evolution* 39 (2000): 57–106.

[19] Norton, Heather L. et al. "Genetic Evidence for the Convergent Evolution of Light Skin in Europeans and East Asians." *Molecular Biology and Evolution* 24 (207): 710–722.

[20] Kraemer, Kenneth H. "Xeroderma Pigmentosum." *Gene Reviews* (2008). http://www.ncbi.nlm.nih.gov/bookshelf/br.fcgi?book=gene&part=xp.

[21] Centers for Disease Control and Prevention. CDC Features. "Cervical Cancer Rates by Race and Ethnicity." http://www.cdc.gov/features/dscervicalcancer/.

[22] Cutts, F. T. et al. "Human Papillomavirus and HPV Vaccines: A Review." *Bulletin of the World Health Organization* 85 (2007): 719–726.

[23] Brentjens, Mathijs H. et al. "Human Papillomavirus: A Review." *Dermatologic Clinics* 20 (2002): 315–331. National Cancer Institute. U.S. National Institutes of Health. "Human Papillomaviruses and Cancer: Questions and Answers." http: //www.cancer.gov/cancertopics/factsheet/Risk/HPV.

[24] Pravda online. "Georgios Papanikolaou the Inventor of the Pap Test." March 2005. http://engforum.pravda.ru/showthread.php?t=225632.

[25] See Weinberg, op. cit., 4, Chapter 3, pp. 57–90 for an excellent account of the tumor virus history.

[26] Temin, Howard M. "The DNA Provirus Hypothesis: The Establishment and Implications of RNA-Directed DNA Synthesis." *Science* 192 (1976): 1075–1080.

[27] Wikipedia. "Virus Classification." http://en.wikipedia.org/wiki/Virus_classification.

[28] Brentjens et al., op. cit., 23. Narisawa-Saito, Mako, and Tohru Kiyono. "Basic Mechanisms of High-Risk Human Papillomavirus-Induced Carcinogenesis: Roles of E6 and E7 Proteins." *Cancer Science* 98 (2007): 1505–1511.

[29] Zielinski, Sarah. Smithsonian.com. "Henrietta Lacks' 'Immortal' Cells." http://www.smithsonianmag.com/science-nature/Henrietta-Lacks-Immortal-Cells.html.

[30] American Cancer Society Statistics. "Breast Cancer." http://www.cancer.org/downloads/STT/F861009_final%209-08-09.pdf.

[31] Petrucelli, Nancie. "BRCA1 and BRCA2 Hereditary Breast/Ovarian Cancer." *Gene Reviews*. http://www.ncbi.nlm.nih.gov/bookshelf/br.fcgi?book=gene&part=brca1.

[32] Schwartz, John. "Cancer Patients Challenge the Patenting of a Gene." *The New York Times*, May 13, 2009.

[33] "Key Phase of Gene-Patent Lawsuit in Judge's Hands." *The Salt Lake Tribune*, February 2, 2010.

[34] ACLU. "ACLU and PUBPAT Argue Today That Patents on Breast Cancer Genes Are Unconstitutional and Invalid." February 10, 2010. http://www.aclu.org/free-speech-womens-rights/aclu-and-pubpat-argue-today-patents-breast-cancer-genes-are-unconstitution. Schwartz, John, and Andrew Pollack. "Judge Invalidates Human Gene Patent." *The New York Times*, March 29, 2010.

[35] Schwartz, op. cit., 32. ACLU, op. cit., 34.

[36] Kepler, Thomas B., Colin Crossman, and Robert Cook-Deegan. "Metastasizing Patent Claims on BRCA1." *Genomics* March 2010 in press. http://www.henrikbranden.se/filer/Material/BRCA-i%20patentartikel.pdf.

[37] National Cancer Institute. "Genetics of Prostate Cancer." http://www.cancer.gov/cancertopics/pdq/genetics/prostate/HealthProfessional/page6.

[38] Ostrander, Elaine A., Kyriacos Markianos, and Janet L. Stanford. "Finding Prostate Cancer Susceptibility Genes." *Annual Review of Human Genetics* 5 (2004): 151–175.

[39] Eeles, Rosalind A. et al. "Multiple Newly Identified Loci Associated with Prostate Cancer Susceptibility." *Nature Genetics* 40 (2008): 316–321. Eeles, Rosalind A. et al. "Identification of Seven New Prostate Cancer Susceptibility Loci Through a Genome-Wide Association Study." *Nature Genetics* 41 (2009): 1116–1121.

[40] National Cancer Institute, op. cit., 37.

[41] "deCODE Prostate Cancer." http://www.decodehealth.com/prostate_cancer.php

[42] Markowitz, Sanford D., and Monica M. Bertagnolli. "Molecular Basis of Colorectal Cancer." *New England Journal of Medicine*. 361 (2009): 2449–2660.

[43] See Bishop, Jerry E., and Michael Waldholz. *Genome*. New York: Simon and Schuster, 1990, pp. 171–178.

[44] Grady, William M. "Genomic Instability and Colon Cancer." *Cancer and Metastasis Reviews* 23 (2004): 11–27.

[45] Genetics Home Reference. "Familial Adenomatous Polyposis." http://ghr.nlm.nih.gov/condition=familialadenomatouspolyposis.

[46] Genetics Home Reference. "APC." http://ghr.nlm.nih.gov/gene=apc.

[47] Genetics Home Reference. "KRAS." http://ghr.nlm.nih.gov/gene=kras.

[48] Genetics Home Reference. "BUB1." http://ghr.nlm.nih.gov/gene=bub1.

[49] Genetics Home Reference. "TGFBR2." http://ghr.nlm.nih.gov/gene=tgfbr2.

[50] Moysich, Kirsten B., Ravi J. Menezes, and Arthur M. Michalek. "Chernobyl-Related Ionizing Radiation Exposure and Cancer Risk: An Epidemiological Review." *The Lancet Oncology* 3 (2002): 269–279.

[51] Proctor, Robert N. "Tobacco and the Global Lung Cancer Epidemic." *Nature Reviews Cancer* 1 (2001): 82–86.

[52] Olstad, Scott. "Cigarette Advertising." *Time*, June 15, 2009.

[53] Proctor, op. cit., 51.

[54] National Cancer Institute. "Lung cancer." http://www.cancer.gov/cancertopics/types/lung.

[55] Hecht, Stephen S. "Tobacco Smoke Carcinogens and Lung Cancer." *Journal of the National Cancer Institute* 91 (1999): 1195–1210.

[56] Office of the Surgeon General. "Secondhand Smoke Is Toxic and Poisonous." http://www.surgeongeneral.gov/library/secondhandsmoke/factsheets/factsheet9.html.

[57] American Cancer Society. "Cigarette Smoking." http://www.cancer.org/docroot/PED/content/PED_10_2X_Cigarette_Smoking_and_Cancer.asp.

[58] Stone, Marvin J. "Thomas Hodgkin: Medical Immortal and Uncompromising Idealist." *BUMC Proceedings* 18 (2005): 368–375.

[59] The Leukemia & Lymphoma Society. "Lymphoma." http://www.leukemia-lymphoma.org/all_page.adp?item_id=7030.

[60] Hartmann, Elena M., German Ott, and Andreas Rosenwald. "Molecular Biology and Genetics of Lymphomas." *Hematology and Oncology Clinics of North America* 22 (2008): 807–823.

[61] Klein, Ulf, and Riccardo Dalla-Favera. "Germinal Centres: Role in B-cell Physiology and Malignancy." *Nature Reviews Immunology* 8 (2008): 22–33.

[62] Genentech. "Rituxan." http://www.gene.com/gene/products/information/oncology/rituxan/.

[63] Non-Hodgkin's Lymphoma Cyberfamily. "Chemotherapy." http://www.nhlcyberfamily.org/treatments/chemotherapy.htm.

[64] HealthInForum. "Pegfilgrastim (Neulasta) from Amgen." http://www.healthinforum.org/Pegfilgrastim-Neulasta-from-Amgen-info-26087.html.

[65] Weinberg, op. cit., 4, pp. 757–765.

[66] The Leukemia & Lymphoma Society. "Chronic Myelogenous Leukemia." http://www.leukemia-lymphoma.org/all_page.adp?item_id=8501.

[67] Vogelstein, Bert, and Kenneth W. Kinzler. "Cancer Genes and the Pathways They Control." *Nature Medicine* 10 (2004): 789–799.

Chapter 6: Genes and behavior

[1] Morrell, Virginia. "Evidence Found for a Possible 'Aggression Gene.'" *Science* 260 (1993): 1722–1723. Han Brunner. ICHG Scientific Program Committee Members. (October 11–15, 2011). http://www.ifhgs.org/pages/ichg_comm_members.shtml.

[2] Brunner, H. G. et al. "X-Linked Borderline Mental Retardation with Prominent Behavioral Disturbance: Phenotype, Genetic Localization, and Evidence for Disturbed Monoamine Metabolism." *American Journal of Human Genetics* 52 (1993): 1032–1039. Brunner H. G. et al. "Abnormal Behavior Associated with a Point Mutation in the Structural Gene for Monoamine Oxidase A." *Science* 262 (1993): 578–580.

[3] Angier, Natalie. "Gene Tie to Male Violence." *The New York Times*, October 22, 1993.

[4] Sabol, Sue Z., Stella Hu, and Dean Hamer. "A Functional Polymorphism in the Monoamine Oxidase A Gene Promoter." *Human Genetics* 103 (1998): 273–279. Deckert, Jürgen et al. "Excess of High Activity Monoamine Oxidase A Gene Promoter Alleles in Female Patients with Panic Disorder." *Human Molecular Genetics* 8 (199): 621–624.

[5] Caspi, Avshalom et al. "Role of Genotype in the Cycle of Violence of Maltreated Children." *Science* 297 (2002): 851–854.

[6] Dunedin Multidisciplinary Health and Development Research Unit. "History of the Study." http://dunedinstudy.otago.ac.nz/about-us/how-we-began/history-of-the-study.

[7] Foley, Debra L. et al. "Childhood Adversity, Monoamine Oxidase A Genotype, and Risk for Conduct Disorder." *Archives of General Psychiatry* 61 (2004): 738–744.

[8] Gibbons, Ann. "Tracking the Evolutionary History of the 'Warrior' Gene." *Science* 304 (2004): 818–819.

[9] Timothy K. Newman et al. "Monoamine Oxidase A Gene Promoter Variation and Rearing Experience Influences Aggressive Behavior in Rhesus Monkeys." *Biological Psychiatry* 57 (2005): 167–172.

[10] Beaver, Kevin et al. "Monoamine Oxidase A Genotype Is Associated with Gang Membership and Weapon Use. *Comprehensive Psychiatry* 51 (2010): 130–134.

[11] Kingsbury, Kathleen. "Which Kids Join Gangs? A Genetic Explanation." *Time*, June 10, 2009.

[12] Hagerty, Barbara Bradley. "Can Your Genes Make You Murder?" National Public Radio. http://www.npr.org/templates/story/story.php?storyId=128043329.

[13] Bernet, William et al. "Bad Nature, Bad Nurture, and Testimony Regarding MAOA and SLC6A4 Genotyping at Murder Trials." *Journal of Forensic Sciences* 52 (2007): 1362–1371.

[14] Hagerty, op. cit., 12.

[15] McDermott, Rose et al. "Monoamine Oxidase A Gene (MAOA) Predicts Behavioral Aggression Following Provocation." *Proceedings of the National Academy of Sciences* 106 (2009): 2118–2123.

[16] De Neve, Jan Emmanuel, and James H. Fowler. "The MAOA Gene Predicts Credit Card Debt." SSRN: Social Science Research Network, abstract, January 27, 2010. http://papers.ssrn.com/sol3/papers.cfm?abstract_id=1457224. The entire paper is also posted on the Web. http://jhfowler.ucsd.edu/maoa_and_credit_card_debt.pdf.

[17] Begley, Sharon. "Millions of Useless Purchases Explained at Last!" *Newsweek*, November 11, 2009. Sager, Ryan. "Overspending: Blame It on Your Debt Gene?" *Smart Money*, November 6, 2009.

[18] "Once Were Warriors: Gene Linked to Maori Violence." *Sydney Morning Herald*, August 9, 2006. Lea, Rod, and Geoffrey Chambers. "Monoamine Oxidase, Addiction, and the 'Warrior' Gene Hypothesis." *The New Zealand Medical Journal* 120 (2007), 1250.

[19] Hook, G. Raumati. "'Warrior genes' and the Disease of Being Maori." *MAI Review* 2 (2009), Target Article.

[20] Caspi, op. cit., 5.

[21] March of Dimes. "Fragile X Syndrome." http://www.marchofdimes.com/professionals/14332_9266.asp.

[22] Martin, J. P., and J. Bell. "A Pedigree of a Mental Defect Showing Sex-Linkage." *Journal of Neurology, Neurosurgery, and Psychiatry* 6 (1943): 154–157.

[23] Harper, Peter S. "Julia Bell and the Treasury of Human Inheritance." *Human Genetics* 116 (2005): 422–432.

[24] Moore, Byron C. "Fragile X-Linked Mental Retardation and Macro-orchidism." *Western Journal of Medicine* 137 (1982): 278–281.

[25] Verkerk, Annemieke J. M. H. et al. "Identification of a Gene (*FMR-1*) Containing a CGG Repeat Coincident with a Breakpoint Cluster Region Exhibiting Length Variation in Fragile X Syndrome." *Cell* 65 (1991): 905–914.

[26] Jewell, Jennifer A. *e*medicine. "Fragile X Syndrome." http://emedicine.medscape.com/article/943776-overview.

[27] Stöger Reinhard et al. "Epigenetic Variation Illustrated by DNA Methylation Patterns of the Fragile-X Gene *FMR1*." *Human Molecular Genetics* 6 (1997): 1791–1801.

[28] Jewell op. cit., 26 The National Fragile X Foundation. "Austism and the Fragile X Syndrome." http://www.fragilex.org/html/autism_and_fragile_x_syndrome.htm. The National Fragile X Foundation. "Characteristics of Cognitive Development." http://www.fragilex.org/html/cognitive.htm.

[29] Bailey, J. Michael, and Richard C. Pillard. "A Genetic Study of Male Sexual Orientation." *Archives of General Psychiatry* 48 (1991): 1089–1096.

[30] LeVay, Simon. "A Difference in Hypothalamic Structure Between Heterosexual and Homosexual Men." *Science* 253 (1991): 1034–1037.

[31] Gelman, David. "Born or Bred?" *Newsweek* 119 (1992) issue 8, p. 46.

[32] President's News Conference, in *Public Papers of the Presidents of the United States, William J. Clinton, 1993, Book 1*, July 19, 1993: published 1994: 1111.

[33] Hamer, Dean H. et al. "A Linkage Between DNA Markers on the X Chromosome and Male Sexual Orientation." *Science* 261 (1993): 321–327.

[34] Maddox, John. "Wilful Public Misunderstanding of Genetics." *Nature* 364 (1993): 281.

[35] Angier, Natalie. "Report Suggests Homosexuality Is Linked to Genes." *New York Times*, July 16, 1993.

[36] Associated Press. "Scientists ID Gene Common to Gays." *Philadelphia Daily News*, July 16, 1993.

[37] "Born Gay?" *Time*, July 26, 1993.

[38] Maugh, Thomas H. II. "Study Strongly Links Genetics, Homosexuality." *Los Angeles Times*, July 16, 1993.

[39] Knox, Richard. "Discovery Prompts Fears of Sexual Manipulation." *Boston Globe*, July 16, 1993.

[40] Fausto-Sterling, Anne, and Evan Balaban. "Genetics and Male Sexual Orientation." *Science* 261 (1993): 1257.

[41] Hamer, Dean et al. "Response." *Science* 261 (1993): 1259.

[42] Risch, Neil, Elizabeth Squires-Wheeler, and Bronya J. B. Keats. "Male Sexual Orientation and Genetic Evidence." *Science* 262 (1993): 2063–2064.

[43] Hamer, Dean, and Peter Copeland. *The Science of Desire*. New York: Simon & Schuster, 1994.

[44] Hu, Stella et al. "Linkage Between Sexual Orientation and Chromosome Xq28 in Males But Not in Females." *Nature Genetics* 11 (1995): 248–256.

[45] Crewdson, John. "Study on 'Gay Gene' Challenged Author Defends Against Allegations." *Chicago Tribune*, June 25, 1995.

[46] Brelis, Matthew. "The Fading Gay Gene." *Boston Globe*, February 17, 1999.

[47] Rice, George et al. "Male Homosexuality: Absence of Linkage to Microsatellite Markers at Xq28." *Science* 284 (1999): 665–667. Wickelgren, Ingrid. "Discovery of 'Gay Gene' Questioned." *Science* 284 (1999): 571.

[48] Mustanski, Brian S. et al. "A Genomewide Scan of Male Sexual Orientation." *Human Genetics* 116 (2005): 272–278.

[49] Martinowich, Keri, Robert J. Schloesser, and Husseini K. Manji. "Bipolar Disorder: From Genes to Behavior Pathways." *Journal of Clinical Investigation* 119 (2009): 726–736.

[50] Risch, Neil, and David Botstein. "A Manic Depressive History." *Nature Genetics* 12 (1996): 351–353.

[51] Hayden, E. P., and J. I. Nurnberger Jr. "Molecular Genetics of Bipolar Disorder." *Genes, Brain and Behavior* 5 (2006): 85–95.

[52] Wikipedia. "Circadian Rhythm." http://en.wikipedia.org/wiki/Circadian_rhythm.

[53] Rivkees, Scott A. "Circadian Rhythms—Genetic Regulation and Clinical Disorders." *Growth, Genetics and Hormones* 18 (2002): 1–6.

[54] Roybal, Kole et al. "Mania-Like Behavior Induced by Disruption of *CLOCK*." *Proceedings of the National Academy of Sciences* 104 (2007): 6406–6411. Coyle, Joseph T. "What Can a Clock Mutation in Mice Tell Us About Bipolar Disorder?" *Proceedings of the National Academy of Sciences* 104 (2007): 6097–6098.

[55] Shi, J. et al. "Clock Genes May Influence Bipolar Disorder Susceptibility and Dysfunctional Circadian Rhythm. *American Journal of Medical Genetics. B. Neuropsychiatric Genetics* 147B (2008): 1047–1055.

[56] Sullivan, Patrick F. "The Genetics of Schizophrenia." *PloS Medicine* 2 (2005): 0614–0618.

[57] St. Clair, David et al. "Association Within a Family of a Balanced Translocation with Major Mental Illness." *The Lancet* 336 (1990): 13–16.

[58] Millar, J. K. et al. "Disruption of Two Novel Genes by a Translocation Cosegregating with Schizophrenia." *Human Molecular Genetics* 9 (2000): 1415–1423.

[59] Millar, J. Kirsty et al. "DISC1 and DISC2: Discovering and Dissecting Molecular Mechanisms Underlying Psychiatric Illness." *Annals of Medicine* 36 (2004): 367–378.

[60] Millar, J. Kirsty et al. "DISC1 and PDE4B Are Interacting Genetic Factors in Schizophrenia That Regulate cAMP Signaling." *Science* 310 (2005): 1187–1191.

[61] Mackie, Shaun, J. Kirsty Millar, and David J. Porteous. "Role of DISC1 in Neural Development and Schizophrenia." *Current Opinion in Neurobiology*. 17 (2007): 95–102.

[62] Sachs, N. A. et al. "A Frameshift Mutation in Disrupted in Schizophrenia 1 in an American Family with Schizophrenia and Schizoaffective Disorder." *Molecular Psychiatry* 10 (2005): 758–764.

[63] Mackie, Millar, and Porteous, op. cit., 61. For example, DISC1 interacts with the NDEL1 and FEZ1 proteins that are involved in development of the nervous system and also with PDE4 that is part of a neurosignaling pathway. The expression of this gene, the interaction of the DISC1 protein with other proteins, and its role in neurogenesis are currently the subjects of intense study in humans and in the mouse model system.

[64] Gillham, Nicholas W. *A Life of Sir Francis Galton: From African Exploration to the Birth of Eugenics*. New York: Oxford University Press, 2001, pp. 340–341.

[65] Nurnberger, John L., Jr., and Laura Jean Eierut. "Seeking the Connections: Alcoholism and Our Genes." *Scientific American* April (2007): 46–53.

[66] Xiao, Q., H. Weiner, and D. W. Crabb. "The Mutation in the Mitochondrial Aldehyde Dehydrogenase (ALDH2) Gene Responsible for Alcohol-Induced Flushing Increases Turnover of the Enzyme Tetramers in a Dominant Fashion." *Journal of Clinical Investigation* 98 (1996): 2027–2032.

[67] Li, Ming D., and Margit Burmeister. "New Insights into the Genetics of Addiction." *Nature Reviews Genetics* 10 (2009): 225–231.

[68] Bierut, Laura J. "A Genome-Wide Association Study of Alcohol Dependence." *Proceedings of the National Academy of Sciences* 107 (2010): 5082–5087.

[69] National Institutes of Health. National Institute on Alcohol Abuse and Alcoholism. "Collaborative Studies on Genetics of Alcoholism (COGA)." http://www.niaaa.nih.gov/ResearchInformation/ExtramuralResearch/SharedResources/projcoga.htm.

[70] Li and Burmeister, op. cit., 67.

Chapter 7: Genes and IQ: an unfinished story

[1] Galton, Francis. "Hereditary Talent and Character." *MacMillan's Magazine* 12 (1865): 157166, 318–327.

[2] Galton, Francis. *Hereditary Genius: An Inquiry into Its Laws and Consequences*. London: Macmillan, 1869.

[3] Gillham, Nicholas Wright. *Sir Francis Galton: From African Exploration to the Birth of Eugenics.* New York: Oxford University Press, 2001.

[4] Human Intelligence. "Alfred Binet: French Psychologist." http://www.indiana.edu/~intell/binet.shtml.

[5] Human Intelligence. "Henry Herbert Goddard: American Psychologist." http://www.indiana.edu/~intell/goddard.shtml.

[6] Vineland Training School History. http://www.vineland.org/history/trainingschool/history/history.html.

[7] For a brief discussion of Cattell's relationship with Galton as well as his own work, see Gillham, op. cit., 3, pp. 227–228.

[8] See the discussion in Kevles, Daniel J. *In the Name of Eugenics*. Cambridge: Harvard University Press, 1995. pp. 77–79.

[9] Human Intelligence. "Lewis Madison Terman: Cognitive Psychologist." http://www.indiana.edu/~intell/terman.shtml. See Kevles, op. cit., 8, pp. 79–81.

[10] Human Intelligence. "Charles Spearman: English Psychologist." http://www.indiana.edu/~intell/spearman.shtml. Williams, Richard H. et al. "Charles Spearman: British Behavioral Scientist." *The Human Nature Review* 3 (2003): 114–118.

[11] Kevles, op. cit., 8, pp. 79–82.

[12] Ibid. pp. 82–83.

[13] Ibid. p. 130.

[14] Goddard, Henry H. "Feeblemindedness: A Question of Definition." *Journal of Psycho Asthenics* 33 (1928): 219–227.

[15] Terman, Lewis. M., and Maud A. Merrill. *Measuring Intelligence: A Guide to the Administration of the New Revised Stanford-Binet Tests of Intelligence*. Boston: Houghton Mifflin, 1937.

[16] Kevles, op. cit., 8, pp. 129–130.

[17] A Brief History of the SAT: Secrets of the SAT. http://www.pbs.org/wgbh/pages/frontline/shows/sats/where/history.html. Piacenza, Matt. "Flawed from the Start: The History of the SAT." http://journalism.nyu.edu/pubzone/race_class/edu-matt3.htm. Blackwell, Jon. "1947: America's Tester-in-Chief." *The Trentonian*. http://www.capitalcentury.com/1947.html.

[18] Fancher, Raymond. *Pioneers of Psychology*, 3rd ed. New York: W. W. Norton & Company, 1996. See esp. chap. 9, "Psychology as the Science of Behavior: Ivan Pavlov, John Watson, and B. F. Skinner." Todd, James T., and Edward K. Morris, ed. *Modern Perspectives on John B. Watson and Classical Behaviorism*. Westport, CT: Greenwood Press, 1994.

[19] Watson, John B. "Psychology as a Behaviorist Views It." *Psychological Review* 20 (1913): 158–177.

[20] Watson, John B., and Rosalie Rayner. "Conditioned Emotional Reactions." *Journal of Experimental Psychology* 3 (1920): 1–14.

[21] Watson, John B. *Behaviorism, revised edition*. Chicago: University of Chicago Press, 1930, p. 82.

[22] Fancher, op. cit., 18 O'Donohue, William, and Kyle E. Ferguson. *The Psychology of B. F. Skinner*. Thousand Oaks, CA: Sage, 2001.

[23] Skinner, B. F. *Walden Two*. New York: Macmillan, 1948.

[24] Skinner, B. F. *Beyond Freedom and Dignity*. New York: Knopf, 1971.

[25] Galton, Francis. "The History of Twins, as a Criterion for the Relative Powers of Nature and Nurture." *Fraser's Magazine* 12 (1875): 566–576.

[26] For example, see Lawrence Wright. "Double Mystery." *The New Yorker*, August 1995, pp. 45–62.

[27] Kevles, op. cit., 8, 140–141.

[28] Human Intelligence. "The Cyril Burt Affair." http://www.indiana.edu/~intell/burtaffair.shtml.

[29] Jensen, Arthur R. "How Much Can We Boost IQ and Scholastic Achievement?" *Harvard Educational Review* 39 (1969): 1–123.

[30] Wikipedia. "How Much Can We Boost IQ and Scholastic Achievement?" http://en.wikipedia.org/wiki/User:David.Kane/How_Much_Can_We_Boost_IQ_and_Scholastic_Achievement%3F.

[31] Jensen, Arthur R. "The Limited Plasticity of Human Intelligence." Originally published in *The Eugenics Bulletin*, Fall 1982. http://forums.skadi.net/archive/index.php/index.php/t-40779.html.

[32] Herrnstein, Richard. "I.Q." *The Atlantic* 228 (1971): 44–64.

[33] Kamin, Leon. *The Science and Politics of IQ*. Potomac, MD: Lawrence Earlbaum Associates, Inc., 1974.

[34] William H. Tucker has written two definitive articles on the controversy. "Re-reconsidering Burt Beyond a Reasonable Doubt." *Journal of the History of the Behavioral Sciences* 33 (1997): 145–162. "Burt's Separated Twins: The Larger Picture." *Journal of the History of the Behavioral Sciences* 43 (2007): 81–86.

[35] Gould, Stephen Jay. *The Mismeasure of Man*. New York: W. W. Norton & Company, 1981.

[36] Lewontin, R. C., Steven Rose, and Leon J. Kamin. *Not In Our Genes: Biology, Ideology, and Human Nature*. New York: Pantheon Books, 1984.

[37] Herrnstein, Richard J., and Charles Murray. *The Bell Curve: Intelligence and Class Structure in American Life*. New York: The Free Press, 1994.

[38] Murray, Charles, and Richard J. Herrnstein. "Race, Genes, and I.Q.—An Apologia." *The New Republic*, October 31, 1994, pp. 27–37.

[39] Fischer, Claude S. et al. *Inequality by Design: Cracking the Bell Curve Myth*. Princeton: Princeton University Press, 1996.

[40] Murray and Herrnstein, op. cit., 38, pp. 307–309.

[41] Flynn, James R. *What Is Intelligence*? New York: Cambridge University Press, 2007.

[42] Mackintosh, N. J. *IQ and Human Intelligence*. New York: Oxford University Press, 1998.

[43] Murray and Herrnstein, op. cit., 38.

[44] Nisbett, Richard E. *Intelligence and How to Get It*. New York: W. W. Norton & Company, 2009.

[45] Leigh Bureau. "Richard Nisbett." http://www.leighbureau.com/speakers/rnisbett/nisbett.pdf.

[46] Parker Martin, Christine. Oscar. "Nature or Nurture: An IQ Study Shows That If You're Poor, It's Not Easy to Bloom Where You Are Planted." http://oscar.virginia.edu/x5701.xml.

[47] Wikipedia. "Kaplan, Inc." http://en.wikipedia.org/wiki/Kaplan,_Inc. Kaplan Test Prep and Admissions http://www.kaptest.com/

[48] Winstein, Keith J. "Kaplan's Tutoring Business Made the Grade." *The Wall Street Journal*, August 25, 2009.

[49] Inlow, Jennifer K., and Linda L. Restifo. "Molecular and Comparative Genetics of Mental Retardation." *Genetics* 166 (2004): 835–881.

[50] Deary, Ian J., W. Johnson, and L.M. Houlihan. "Genetic Foundations of Human Intelligence." *Human Genetics* 126 (2009): 215–232.

[51] Newton, Giles. The Human Genome. "Genes and Cognition." http://genome.wellcome.ac.uk/doc_WTD020801.html.

[52] Duncan, David Ewing. "On a Mission to Sequence the Genomes of 100,000 People." *The New York Times*, Tuesday, June 8, 2010, D3.

Chapter 8: Preventing genetic disease

[1] Katz, Linda G. "*Howard v. Lecher*: An Unreasonable Limitation on a Physician's Liability in a Wrongful Life Suit." *New England Law Review* 12 (1977): 819–840. Pelias, Mary Z. "Torts of Wrongful Birth and Wrongful Life: A Review." *American Journal of Medical Genetics* 25 (1986): 71–80.

[2] Annas, George. "Medical Paternity and Wrongful Life." *The Hastings Center Report* 9 (1979): 15–17. Pelias, op. cit., 1 Smith, Don C., and Joseph D. McInerney. "Human Genetics and the Law: Biological Progress and Challenges to Traditional Values." *The American Biology Teacher* 47 (1985): 396–401. Culliton, Barbara. "Briefing: Physicians Sued for Failing to Give Genetic Counseling." *Science* 203 (1979): 251.

[3] Nelson, Merrill F. "Wrongful Life—Impaired Infant's Cause of Action Recognized: *Curlender v. Bio-Science Laboratories*." lawreview.byu.edu/archives/1980/3/nel.pdf. Kevles, Daniel. *In the Name of Eugenics*. Cambridge: Harvard University Press, 1985, p. 293. Curlender v. Bio-Science Laboratories (1980) 106 Cal.App.3d 911, 165 Cal.Rptr. 477 http://www.lawlink.com/research/CaseLevel3/56757.

[4] Kevles, Daniel J. *In the Name of Eugenics*. Cambridge: Harvard University Press, 1995. p. 253. Gorski, Jermoe L. "James V. Neel, 1915–2000: In Memorium." *American Journal of Medical Genetics* 95 (2000): 1–3. Dice, Lee R. "Heredity Clinics: Their Value for Public Service and for Research." *The American Journal of Human Genetics* 4 (1952): 1–13.

[5] Kevles, op. cit., 4, p. 253. The Dight Institute of the University of Minnesota. *The Journal of Heredity* 35 (1944): 32. Phelps, Gary. "The Eugenics Crusade of Charles Fremont Dight." *Minnesota History* Fall 1984, pp. 99–108.

[6] Kevles, Ibid. Resta, R. G. "The Historical Perspective: Sheldon Reed and 50 Years of Genetic Counseling." *Journal of Genetic Counseling* 6 (1997): 375–377. Anderson, V. Elving. "Obituary: Sheldon C. Reed, Ph.D. (November 7, 1910–February 1, 2003): Genetic Counseling, Behavioral Genetics." *American Journal of Human Genetics* 73 (2003): 1–4.

[7] Begleiter, Michael L. "Training for Genetic Counselors." *Nature Reviews Genetics* 3 (2002): 557–561. Bennett, Robin L. et al. "Genetic Counselors: Translating Genomic Science into Practice." *The Journal of Clinical Investigation*. 112 (2003): 1274–1279. Sarah Lawrence College. "Human Genetics Program History, 1972–1980." http://www.slc.edu/graduate/programs/human-genetics/character/history/1972-1980.html Motulsky, Arno G. "2003 ASHG Award for Excellence in Human Genetics Education: Introductory Speech for Joan Marks." *American Journal of Human Genetics* 74 (2004): 393–394. Kenen, Regina H. "Genetic Counseling: The Development of a New Interdisciplinary Occupational Field." *Social Science and Medicine* 18 (1984): 541–549.

[8] Begleiter, op. cit., 7.

[9] American Board of Genetic Counseling: General Information. http://www.abgc.net/english/View.asp?x=1465.

[10] Bennett, Robin L. et al. op. cit.,7. Ciarleglio, Leslie J. et al. "Genetic Counseling Throughout the Life Cycle." *The Journal of Clinical Investigation* 112 (2003): 1280–1285.

[11] A search of Genetics Home Reference yielded 172 entries.

[12] Leshin, Len. "Prenatal Screening for Down Syndrome." http://www.ds-health.com/prenatal.htm. Newberger, David S. "Down Syndrome: Prenatal Risk Assessment and Diagnosis." *American Family Physician,* August 15, 2000. http://www.aafp.org/afp/20000815/825.html

[13] Barstow, David. "An Abortion Battle, Fought to the Death." *The New York Times*, July 26, 2009. "Genetic Counselor Blasts National Society of Genetic Counselors For Not Putting Out a Public Statement Condemning Dr. Tiller's Murder." *Democracy Now*, June 10, 2009. "National Society of Genetic Counselors Releases Statement on Murder of Dr. George Tiller." http://www.newsguide.us/health-medical/general/National-Society-of-Genetic-Counselors-Releases-Statement-on-Murder-of-Dr-George-Tiller/?date=2009-06-14. Hegeman, Roxana. "Scott Roeder Sentenced to Life in Prison: George Tiller's Murderer Describes Abortion In Court." *The Huffington Post*, July 6, 2010. http://www.huffingtonpost.com/2010/04/01/scott-roeder-sentenced-to_n_522654.html?view=print. Stumpf, Joe, and Monica Davey. "Abortion Doctor Shot to Death in Kansas Church." *The New York Times*, June 1, 2009.

[14] Bennett, Robin L. et al., op. cit., 7. Ciarleglio, Leslie J. et al., op. cit., 10.

[15] Westen, John-Henry. "Pope Condemns In Vitro Fertilization: 'Barrier protecting human dignity has been broken.'" LifeSiteNews.com, January 31 2008. http://www.lifesitenews.com/ldn/printerfriendly.html?articleid=08013110.

[16] "World's First Test-Tube Baby Is Pregnant." *Mail Online,* 10 July 2006. http://www.dailymail.co.uk/news/article-394894/Worlds-test-tube-baby-pregnant.html?printingPage=true. BBC News. "Baby Son Joy for Test-Tube Mother." http://news.bbc.co.uk/2/hi/uk_news/6260171.stm.

[17] Biggers, J. D. "Walter Heape, FRS: A Pioneer in Reproductive Biology. Centenary of His Embryo Transfer Experiments." *Journal of Reproduction and Fertility* 93 (1991): 173–186. Morris, Randy S. *IVF1.* "IVF—In Vitro Fertilization." http://www.ivf1.com/ivf/. World In Vitro Fertilization Units. "The History of IVF—Major Milestones of the Process." http://www.ivf-programm.de/IVF-History.pdf. Kevles, Daniel., op. cit., 4, p. 189.

[18] Edwards, Robert G. "The Bumpy Road to Human In Vitro Fertilization." *Nature Medicine* 7 (2001): 1091–1094. Golden, Frederic. "Patrick Steptoe and Robert Edwards: Brave New Baby Doctors." *Time,* Monday March 29, 1999. Spar, Debora L. *The Baby Business*. Cambridge: Harvard Business School Press, 2006, pp. 21–26. Cohen, Jean C. et al. "The Early Days of IVF Outside the UK." *Human Reproduction Update* 11 (2005): 439–459.

[19] This quotation and the ones in the following paragraph are taken from Spar Ibid. p. 26. Spar's book gives an excellent account of the economics of what she calls "The Baby Business."

[20] Ibid. pp. 26–30.

[21] Centers for Disease Control and Prevention. "Assisted Reproductive Technology (ART)." http://www.cdc.gov/art/. Centers for Disease Control and Prevention. "2007 ART Report Section 1-Overview CDC." http://www.cdc.gov/art/ART2007/section1.htm#f3. Centers for Disease Control and Prevention. "2007 ART Report Section 5-ART Trends 1998–2007." http://www.cdc.gov/art/ART2007/section5.htm.

[22] Ibid. Human Fertilsation and Embryology Authority. http://www.hfea.gov.uk/135.html.

[23] Roberts, MacKenna. IVF.net. "UK Parliament Alarmed by 1.2 Million Leftover Embryos." January 9, 2008. http://www.ivf.net/ivf/uk_parliament_alarmed_by_1_2_million_leftover_ivf_embryos-o3162.html.

[24] "What Is SART?" http://www.sart.org/WhatIsSART.html. Hoffman, David I. et al. "Cryopreserved Embryos in the United States and Their Availability for Research." *Fertility and Sterility* 79 (2003): 1063–1069.

[25] Hoffman, David I. et al. "Cryopreserved Embryos in the United States and Their Availability for Research." *Fertility and Sterility* 79 (2003): 1063–1069.

[26] Gurmankin, Andrea D., Sisti, Dominic, and Arthur L. Caplan. "Embryo Disposal Practices in IVF Clinics in the United States." *Politics and the Life Sciences* 22 (2004): 4–8.

[27] Nachtigall, Robert D. et al. "Parents' Conceptualization of Their Frozen Embryos Complicates the Disposition Decision." *Fertility and Sterility* 84 (2005): 431–434.

[28] MercoPress. "Catholic Church Insists in Questioning Nobel Prize to In Vitro fertilization." http://en.mercopress.com/2010/10/05/catholic-church-insists-in-questioning-nobel-prize-to-in-vitro-fertilization.

[29] Edwards, R. G., and R. L. Gardner. "Sexing of Live Rabbit Blastocysts." *Nature* 214 (1967): 576–577.

[30] Handyside, A. H. et al. "Biopsy of Human Preimplantation Embryos and Sexing by DNA Amplification." *The Lancet* 333 (1989): 347–349. Handyside, A. H. et al. "Pregnancies from Biopsied Human Preimplantation Embryos Sexed by Y-specific DNA Amplification." *Nature* 344 (1990): 768–770.

[31] Handyside, A. H. et al. "Birth of a Normal Girl After In Vitro Fertilization and Preimplantation Diagnostic Testing for Cystic Fibrosis." *New England Journal of Medicine* 327 (1992): 951–953.

[32] Gessen, Masha. *Blood Matters: From Inherited Illness to Designer Babies How the World and I Found Ourselves in the Future of the Gene*. Orlando: Houghton Mifflin Harcourt Publishers, 2008, Chapter 12. Hevesi, Dennis. "Yury Verlinsky, Expert in Embryonic Screening, Is Dead at 65." *The New York Times*, July 23, 2009. Maugh, Thomas H. "Yury Verlinsky; Russian émigré was Pioneer in Prenatal Testing." *Los Angeles Times*, July 23, 2009. Morris, Randy S. IVF1 "Remembering Yury Verlinsky, PhD." http://www.ivf1.com/yury-verlinsky/. Wikipedia. "Yury Verlinski." http://en.wikipedia.org/wiki/Yury_Verlinsky.

[33] Reproductive Genetics Institute. http://www.reproductivegenetics.com/.

[34] Spar, op. cit., 18, pp. 117–118. Genesis Genetics Institute. http://www.genesisgenetics.org/.

[35] Genetics & IVF Institute. http://www.givf.com/.

[36] Reproductive Science Center. "Insurance." http://www.rscnewengland.com/infertility-insurance.html.

Chapter 9: Treating genetic disease

[1] Who Named It? "Robert Guthrie." http://www.whonamedit.com/doctor.cfm/3435.html. Bjorhus, Jennifer. "Dr. Robert Guthrie, Developed Pku Test." *The Seattle Times*, June 28, 1995. http://community.seattletimes.nwsource.com/archive/?date=19950628&slug=2128681. "Obituaries: Robert Guthrie, Professor; Developer of PKU Test." *Reporter*, August 31, 1995. http://www.buffalo.edu/ubreporter/archives/vol27/vol27n01/n14.html.

[2] Guthrie, Robert, and Ada Susi. "A Simple Phenylalanine Method for Detecting Phenylketoneuria in Large Populations of Newborn Infants." *Pediatrics* 32 (1963): 338–343.

[3] Kevles, Daniel. *In the Name of Eugenics*. Cambridge: Harvard University Press, 1985, p. 255.

[4] Brody, Jane E. "PERSONAL HEALTH; Pounds of Prevention in Drops of Baby Blood." *The New York Times*, January 29, 2002.

[5] "Newborn Screening: Toward a Uniform Screening Panel and System." *Federal Register* 70, no. 44, March 8, 2005. Kolata, Gina. "Panel to Advise Testing Babies for 29 Diseases." *The New York Times*, February 21, 2005.

[6] "National Newborn Screening Status Report." Updated 03/01/10. http://genes-r-us.uthscsa.edu/nbsdisorders.pdf.

[7] "The Changing Moral Focus of Newborn Screening: An Ethical Analysis by the President's Council on Bioethics." Washington, DC, December 2008. http://www.bioethics.gov.

[8] McElroy, Molly. AAAS News: News Archives. "Researchers and Policymakers Point to Success and Challenges in Personalized Medicine." http://www.aaas.org/news/releases/2009/1214personalized_medicine.shtml.

[9] "Blood from Newborn Tests Stir Ethics Debate." *Daily Herald*, February 15, 2010. http://www.dailyherald.com/story/?id=358327.

[10] Driver, Carol. "DNA from Millions of Newborn Babies Is Secretly Stored on NHS Database." *Mail Online*. http://www.dailymail.co.uk/news/article-1280891/NHS-creates-secret-database-babies-blood-samples-parental-consent.html?printingPage=true.

[11] Michigan Bio Trust For Health: Questions and Answers. http://www.michigan.gov/mdch/0,1607,7-132-2942_4911_4916_53246-232933—,00.html#How_is_privacy_protected?

[12] "In Memoriam: Dr. Menkes Was Expert on Inherited Metabolic Disease." AAP News. http://aapnews.aappublications.org/cgi/reprint/30/2/25. Maugh, Thomas H. II. "John Menkes, 79; Neurologist Identified Congenital Disorders." http://www.boston.com/bostonglobe/obituaries/articles/2008/12/05/john_menkes_79_neurologist_identified_congenital_disorders/. Menkes John H. et al. "A New Syndrome: Progressive Familial Infantile Cerebral Dysfunction Associated with an Unusual Urinary Substance. *Pediatrics* 14 (1954): 462–467.

[13] Dent, C. E., and R. G. Westall. "Studies in Maple Syrup Urine Disease." *Archives of Disease in Childhood* 36 (1961): 259–268.

[14] Naylor, Edwin W., and Robert Guthrie. "Newborn Screening for Maple Syrup Urine Disease (Branched-Chain Ketoaciduria)." *Pediatrics* 61 (1978): 262–266.

[15] Strauss, Kevin A., Erik G. Puffenberger, and D. Holmes Morton. "Maple Syrup Urine Disease." Last Update: December 15, 2009. *Gene Reviews*. http://www.ncbi.nlm.nih.gov/bookshelf/br.fcgi?book=gene&part=msud.

[16] King, Lisa Sniderman, Cristine Trahms, and C. Ronald Scott. "Tyrosinemia Type I." *Gene Reviews*. http://www.ncbi.nlm.nih.gov/bookshelf/br.fcgi?book=gene&part=tyrosinemia#tyrosinemia.Management.

[17] Genetics Home Reference. "Congenital Hypothyroidism." http://ghr.nlm.nih.gov/condition/congenital-hypothyroidism. Rastogi, Maynika V., and Stephen H. LaFranchi. "Congenital Hypothyroidism." *Orphanet Journal of Rare Diseases* 5 (2010): 17–39. http://www.ojrd.com/content/5/1/17.

[18] Genetics Home Reference. "21-hydroxylase Deficiency." http://ghr.nlm.nih.gov/condition/21-hydroxylase-deficiency/show/print. Google Health. "Congenital Adrenal Hyperplasia." https://health.google.com/health/ref/Congenital+adrenal+hyperplasia. Wilson, Thomas A. *e*medicine. "Congenital Adrenal Hyperplasia." http://emedicine.medscape.com/article/919218-print.

[19] World Federation of Hemophilia. "Guidelines for the Management of Hemophilia." http://www.ehc.eu/fileadmin/dokumente/Gudelines_Mng_Hemophilia.pdf. Schwartz, Robert A., Elzbieta Klujszo, and Rajalaxmi McKenna. *e*medicine. "Factor VIII." Updated 8/13/09. Genetics Home Reference. "Hemophilia." http://ghr.nlm.nih.gov/condition/hemophilia.

[20] Bogdanich, Walt, and Eric Koli. "2 Paths of Bayer Drug in 80's: Riskier One Steered Overseas." *The New York Times*, May 22, 2003. http://www.nytimes.com/ 1996/06/11/business/blood-money-aids-hemophiliacs-are-split-liability-cases-bogged-down-disputes.html?pagewanted=print. http://emedicine.medscape.com/article/201319-overview. Contaminated haemophilia blood products. Wikipedia. http://emedicine.medscape.com/article/201319-overview. HIV/AIDS. National Hemophilia Foundation. http://www.hemophilia.org/ NHFWeb/MainPgs/MainNHF.aspx?menuid=43&contentid=39&rptname=bloodsafety. Meier, Barry. "Blood, Money and AIDS: Hemophiliacs Are Split: Liability Cases Bogged Down in Disputes." *The New York Times*, June 11, 1996. http://partners.nytimes.com/library/national/science/aids/061196sci-aids.html.

[21] Bogdanich and Koli, op. cit., 20.

[22] "World: Europe, Blood Scandal Ministers Walk Free." *BBC News*, March 9, 1999. http://news.bbc.co.uk/2/hi/europe/293367.stm. Breo, Dennis L. "Blood, Money, and Hemophiliacs—The Fatal Story of France's 'AIDSgate.'" *Journal of the American Medical Association* 266 (1991): 3477–3482. Dorfman, Andrea. "Bad Blood in France." *Time*, July 8, 1991. Hunter, Mark. "Blood Money." *Discover*, August 1, 1993. Ingram, Mike. "Court Acquits Former Prime Minister." *World Socialist Web Site*, March 12, 1999. http://www.wsws.org/articles/1999/mar1999/hiv-m12.shtml. Wikipedia. "Infected Blood Scandal (France)." http://en.wikipedia.org/wiki/Infected_blood_scandal_(France).

[23] Bogdanich and Koli, op. cit., 20.

[24] Mannucci, Pier M., and Edward G. D. Tuddenham. "The Hemophilias—From Royal Genes to Gene Therapy." *The New England Journal of Medicine* 344 (2001): 1773–1779.

[25] Recombinant Factor VIII Comparison Chart. http://www.hemophilia.ca/files/rFVIII%20Comparison%20Chart%20-%2001-03-2010.pdf.

[26] Manco-Johnson, Marilyn J. et al. "Prophylaxis Versus Episodic Treatment to Prevent Joint Disease in Boys with Severe Hemophilia." *New England Journal of Medicine* 357 (2007): 535–544. Roosendaal, Goris, and Floris Lafeber. "Prophylactic Treatment for Prevention of Joint Disease in Hemophilia—Cost Versus Benefit." *New England Journal of Medicine* 357 (2007): 603–605.

[27] National Hemophilia Foundation. "Financial and Insurance Issues." http://www.hemophilia.org/NHFWeb/MainPgs/MainNHF.aspx?menuid=34&contentid=24.

[28] Grabowski, Gregory A., and Robert J. Hopkin. "Enzyme Therapy for Lysosomal Storage Disease: Principles, Practice, and Prospects." *Annual Review of Genomics and Human Genetics* 4 (2003): 403–436. Lim-Melia, Elizabeth R., and David F. Kronn. "Current Enzyme Replacement Therapy for the Treatment of Lysosomal Storage Diseases." *Pediatric Annals* 38 (2009): 448–455. Wikipedia. "Macrophage."

http://en.wikipedia.org/wiki/Macrophage. Wikipedia. "Mannose Receptor." http://en.wikipedia.org/wiki/Mannose_receptor.

[29] Beutler, Ernest. "Lysosomal Storage Diseases: Natural History and Ethical and Economic Aspects." *Molecular Genetics and Metabolism* 88 (2006): 208–215.

[30] Office of Inspector General: Health and Human Services. "The Orphan Drug Act: Implementation and Impact," May 2001. http://oig.hhs.gov/oei/reports/oei-09-00-00380.pdf.

[31] Anand, Geeta. "How Drugs for Rare Diseases Became Lifeline for Companies." *The Wall Street Journal*, November 15, 2005. Roscoe Brady M.D. Childrens Gaucher Research Fund. http://www.childrensgaucher.org/about-2/advi. "Dr. Roscoe Brady & Gaucher Disease." http://history.nih.gov/exhibits/gaucher/docs/page_02.html. Federal and Private Roles in the Development of Alglucerase Therapy for Gaucher Disease. OTA-BP-H-104 October 1992. Gagnon, Geoffrey. "So, This is What a Biotech Tycoon Looks Like." *Boston Magazine*, May 19, 2008.

[32] Feuerstein, Adam. "Genzyme's Termeer: Worst Biotech CEO of '09." *The Street*. http://www.thestreet.com/story/10627877/genzymes-termeer-worst-biotech-ceo-of-09.html.

[33] "Shire Announces FDA Approval of VPRIV™ (Velaglucerase Alfa for Injection) for the Treatment of Type 1 Gaucher Disease." *Shire*, February 26, 2010. http://www.shire.com/shireplc/en/investors/.../irshirenews?id=331. "UPDATE2-U.S. Approves Shire's Drug for Gaucher Disease." *Reuters*, February 26, 2010. http://www.reuters.com/article/idUSN2620418520100226. "EU Approves Shire's Vpriv for Gaucher Disease." *EuroPharma Today*, September 7, 2010. http://www.europharmatoday.com/2010/09/eu-approves-shires-vpriv.html.

[34] McBride, Ryan. "Sanofi-Aventis Launches a Hostile Takeover Bid for Genzyme." Xconomy: Boston. October 4, 2010. http://www.xconomy.com/boston/2010/10/04/sanofi-aventis-launches-hostile-takeover-bid-for-genzyme/.

[35] Lim-Melia and Kronn, op. cit., 28. Pastores, Gregory M. "Miglustat: Substrate Reduction Therapy for Lysosomal Storage Disorders Associated with Primary Nervous System Involvement." *Recent Patents on CNS Drug Discovery* 1 (2006): 77–82.

[36] Genetics Home Reference. "Adenosine Deaminase Deficiency." http://ghr.nlm.nih.gov/condition/adenosine-deaminase-deficiency. Hershfield, Michael. "Adenosine Deaminase Deficiency." *Gene Reviews*. Last revision: July 14, 2009. http://www.ncbi.nlm.nih.gov/bookshelf/br.fcgi?book=gene&part=ada#ada.Management.

[37] Chan, Belinda et al. "Long-Term Efficacy of Enzyme Replacement Therapy for Adenosine Daminase (ADA)-Deficient Severe Combined Immunodeficiency (SCID). *Clinical Immunology* 117 (2005): 133–143.

[38] McVicker, Steve. "Bursting the Bubble." *Houston Press*, April 10, 1997. http://www.houstonpress.com/1997-04-10/news/bursting-the-bubble/. Wikipedia. "David Vetter." http://en.wikipedia.org/wiki/David_Vetter.

[39] Naam, Ramez. *More Than Human*. New York: Broadway Books, 2005, Chapter 1. "ADA (Adenosine Deaminase) Deficiency—Treatments for ADA Deficiency." http://science.jrank.org/pages/66/ADA-Adenosine-Deaminase-Deficiency.html.

[40] Blaese, R. Michael et al. "T Lymphocyte-Directed Gene Therapy for ADA-SCID: Initial Trial Results After 4 Years." *Science* 270 (1995): 475–480. Muul, Linda Mesler et al. "Persistence and Expression of the Adenosine Deaminase Gene for 12 Years and Immune Reaction to Gene Transfer Components: Long-Term Results of the First Clinical Gene Therapy Trial." *Blood* 101 (2003): 2563–2569.

[41] Raper, Steven E. et al. "A Pilot Study of In Vivo Liver-Directed Gene Transfer with an Adenoviral Vector in Partial Ornithine Transcarbamylase Deficiency." *Human Gene Therapy* 13 (2002): 163–175. Raper, Steven E. et al. "Fatal Systemic Inflammatory Response Syndrome in a Ornithine Transcarbamylase Deficient Patient Following Adenoviral Gene Transfer." *Molecular Genetics and Metabolism* 80 (2003): 148–158. Wilson, James M. "Lessons Learned from the Gene Therapy Trial for Ornithine Transcarbamylase Deficiency." *Molecular Genetics and Metabolism* 96 (2009): 151–157.

[42] Stolberg, Sheryl Gay. "The Biotech Death of Jesse Gelsinger." *The New York Times*, November 28, 1999.

[43] Stolberg, Sheryl Gay. "F.D.A. Officials Fault Penn Team in Gene Therapy Death." *The New York Times*, December 9, 1999. Stolberg, Sheryl Gay. "Gene Therapy Ordered Halted at University." *The New York Times*, January 22, 2000. Stolberg, Sheryl Gay. "Senators Press for Answers on Gene Trials." *The New York Times*, February 3, 2000. National News Briefs. "Suit Filed Over Death in Gene Therapy Test." *The New York Times*, September 19, 2000. National News Briefs. "Penn Settles Suit on Genetic Test." *The New York Times*, November 4, 2000.

[44] Cavazzana-Calvo, Marina et al. "Gene Therapy of Human Severe Combined Immunodeficiency (SCID)-X1 Disease." *Science* 288 (2000): 669–672. Cavazzana-Calvo, Marina et al. "Gene Therapy for Severe Combined Immunodeficiency." *Annual Review of Medicine* 56 (2005): 585–602.

[45] Connor, Steve. "Doctors Claim a First Genetic 'Cure' for Rhys, the Boy in the Bubble." *The Independent*, April 4, 2002. "Gene Therapy Provides Cure for 'Bubble Baby' Rhys." *Mail Online*. http://www.dailymail.co.uk/news. "Gene Therapy Saves Life of Boy in Bubble." *The Daily Telegraph*, April 4, 2002.

[46] Hacein-Bey-Abina, Salima et al. "Insertional Oncogenesis in 4 Patients After Retrovirus-Mediated Gene Therapy of SCID-X1." *The Journal of Clinical Investigation* 118 (2008): 3122–3142. Henderson, Mark. "Junk Medicine: Pioneers Cannot Put Safety First." *Times Online*, October 11, 2003. http://www.timesonline.co.uk/tol/life_and_style/health/article1032989.ece. Highfield, Roger. "Gene Transplant 'Cancer Risk.'" *Telegraph*, April 27, 2006. http://www.telegraph.co.uk/news/worldnews/europe/france/1516758/Gene-transplant-cancer-risk.html. "Why Gene Therapy Caused Leukemia In Some 'Boy in the Bubble Syndrome' Patients." *Science Daily*, August 10, 2008. http://www.sciencedaily.com/releases/2008/08/080807175438.htm.

[47] Pike-Overzet, Karin et al. "Is IL2RG Oncogenic in T-Cell Development?" *Nature* 443 (2006): E5. Thrasher, Adrian J. et al. "X-SCID Transgene Leukaemogenicity." *Nature* 443 (2006): E5–E6. Woods, Niels-Bjarne et al. "Therapeutic Gene Causing Lymphoma." *Nature* 440 (2006): 1123. Woods et al. "Reply." *Nature* 443 (2006): E6–E7.

[48] Check, Erika. "Gene Therapy Put on Hold as Third Child Develops Cancer." *Nature* 433 (2005): 561. Check, Erika. "Gene-Therapy Trials to Restart Following Cancer Risk Review." *Nature* 434 (2005): 127. Marcus, Amy Dockser. "Study Shows Hope for Gene Therapy." *The Wall Street Journal*, July 10, 2010. "International Gene Therapy Trial Launched at Children's Hospital Boston for 'Bubble Boy' Syndrome." Children's Hospital Boston. July 10, 2010. http://www.childrenshospital.org/newsroom/Site1339/mainpageS1339P1sublevel644.html. Stolberg, Sheryl Gay. "Trials Are Halted on Gene Therapy." *The New York Times*, October 4, 2002.

[49] Aliuti, Alessandro et al. "Correction of ADA-SCID by Stem Cell Gene Therapy Combined with Nonmyeloablative Conditioning." *Science* 296 (2002): 2410–2413. Aiuti, Alessandro, and Maria Grazia Roncarlo. "Ten Years of Gene Therapy for Primary Immune Deficiencies." *Hematology* (2009): 682–689. Aiuti, Alessandro et al. "Gene Therapy for Immunodeficiency Due to Adenosine Deaminase Deficiency." *The New England Journal of Medicine* 360 (2009): 447–458.

[50] Belluck, Pam. "Giving Sight By Therapy With Genes." *The New York Times*, November 3, 2009. Johnson, Carolyn Y. "Genetic Therapy Gets Closer to a 'Cure.'" *The Boston Globe*, February 29, 2009. Kaiser, Jocelyn. "Gene Therapy Helps Blind Children See." *Science Now*, October 24, 2009. http://news.sciencemag.org/sciencenow/2009/10/24-01.html. Maguire, Albert et al. "Age-Dependent Effects of *RPE65* Gene Therapy for Leber's Congenital Amaurosis: A Phase 1 Dose-Escalation Trial." *The Lancet* 374 (2009): 1597–1605. RPE65 Genetics Home Reference. http://ghr.nlm.nih.gov/gene/RPE65.

[51] Carolyn Johnson, op. cit., 50.

[52] Fischer, Alain, and Marina Cavazzana-Calvo. "Gene Therapy of Inherited Diseases." *The Lancet* 371 (2008): 2044–2047. O'Connor, Timothy P., and Ronald G. Crystal. "Genetic Medicines: Treatment Strategies for Hereditary Disorders." *Nature Reviews Genetics* 7 (2006): 261–276. Kimmelman, Jonathan. "The Ethics of Human Gene Transfer." *Nature Reviews Genetics* 9 (2008): 239–244.

[53] Wilson, James M. "A History Lesson for Stem Cells." *Science* 324 (2009): 727–728.

[54] Fox, Maggie. "First Patient Treated in Geron Stem Cell Trials." *Reuters*. http://www.reuters.com/article/idUSTRE69A27F20101011. Pollack, Andrew. "F.D.A. Approves a Stem Cell Trial." *The New York Times*, January 23, 2009.

[55] Alter, Blanche P. eMedicine Pediatrics: General Medicine. "Anemia, Fanconi." http://emedicine.medscape.com/article/960401-overview. Taniguchi, Toshiyasu. "Fanconi Anemia." *Gene Reviews*. http://www.ncbi.nlm.nih.gov/bookshelf/br.fcgi?book=gene&part=fa.

[56] Belkin, Lisa. "The Made-to-Order Savior." *The New York Times*, July 1, 2001. Mundy, Liza. "Life in Henry's Wake: Laurie Strongin Won't Let Her Son's Death Be a Total Loss." *Washington Post*, March 25, 2010. Verlinsky, L. Yuri et al. "Preimplantation Diagnosis for Fanconi Anemia Combined with HLA Matching." *JAMA* 285 (2001): 3130–3133.

Chapter 10: The dawn of personalized medicine

[1] Cystic Fibrosis Foundation. "FAQs About VX-770." http://www.cff.org/research/ClinicalResearch/FAQs/VX-770/. Finder, Jonathan. "In the Pipeline: New Treatments for Cystic Fibrosis." *RT Magazine*. http://www.rtmagazine.com/issues/articles/2010-02_04.asp. Van Goor, Frederick et al. "Rescue of CF Airway Epithelial Cell Function In Vitro by a CFTR Potentiator, VX-770." *Proceedings of the National Academy of Sciences of the United States*. 106 (2009): 18825–18830. VX-770 Vertex. http://www.vrtx.com/current-projects/drug-candidates/vx-770.html. "Vertex Reviews 2010 Business Priorities to Support Goal of Becoming Fully-Capable Biopharmaceutical Company." http://files.shareholder.com/downloads/VRTX/0x0x343162/2120270d-e491-4470-b3d2-c040e7b2532d/VRTX_News_2010_1_10_Webcasts.pdf.

[2] Groopman, Jerome. "Open Channels; Annals of Medicine." *The New Yorker* 85 (2009): 30–36.

[3] Cystic Fibrosis Foundation. "CF Investigational Drug VX-809 Shows Encouraging Results in Phase 2a Trial." http://www.cff.org/aboutCFFoundation/NewsEvents/02-03-VX-809-Shows-Encouraging-Results-in-Phase2a.cfm. VX-809 Vertex. http://www.vrtx.com/current-projects/drug-candidates/VX-809.html. "Vertex Announces Results from Phase 2a Trial of VX-809 Targeting the Defective Protein Responsible for Cystic Fibrosis." Vertex. http://investor.shareholder.com/vrtx/releasedetail.cfm?ReleaseID=442429. "Vertex Reviews 2010 Business Priorities," see op. cit., 1.

[4] Chiaw, Patrick Kim et al. "A Chemical Corrector Modifies the Channel Function of F508del-CFTR." *Molecular Pharmacology* 78 (2010): 411–418. Robert, Renaud et al. "Correction of the ΔPhe 508 Cystic Fibrosis Transmembrane Conductance Regulator Trafficking Defect by the Bioavailable Compound Glafenine." *Molecular Pharmacology* 77 (2010): 922–930. Loo, Tip W., M. Claire Bartlett, and David M. Clarke. "Correctors Promote Folding of the CFTR in the Endoplasmic Reticulum." *Biochemical Journal* 413 (2008): 29–36.

[5] PTC Therapeutics. "Ataluren for Genetic Disorders." http://www.ptcbio.com/3.1.1_genetic_disorders.aspx. Karem, Eitan et al. "Effectiveness of PTC124 Treatment of Cystic Fibrosis Caused by Nonsense Mutations: A Prospective Phase II Trial." *The Lancet* 372 (2008): 719–727. Welch, Ellen M. et al. "PTC124 Targets Genetic Disorders Caused by Nonsense Mutations." *Nature* 447 (2007): 87–91.

[6] Lazarou, J., B. H. Pomeranz, and P. N. Corey. "Incidence of Adverse Drug Reactions in Hospitalized Patients: A Meta-Analysis of Prospective Studies." *JAMA* 279 (1998): 1200–1205. Pharmacogenomics. http://www.ornl.gov/sci/techresources/Human_Genome/medicine/pharma.shtml.

[7] Beals, Jacquelyn K. "FDA Releases List of Genomic Biomarkers Predictive of Drug Interactions." *Medscape Medical News*, August 1, 2008. http://www.medscape.com/viewarticle/578464. "Table of Valid Genomic Biomarkers in the Context of Approved Drug Labels." http://www.fda.gov/Drugs/ScienceResearch/ResearchAreas/Pharmacogenetics/ucm083378.htm.

[8] The Body: The Complete HIV/AIDS Resource. "Co-Receptors: CCR5—Understanding HIV." http://www.thebody.com/content/art4978.html?ts=pf. "SELZENTRY (maraviroc) tablets." http://www.selzentry.com/.

[9] Moyer, Thomas P. et al. "Warfarin Sensitivity Genotyping: A Review of the Literature and Summary of Patient Experience." *Mayo Clinic Proceedings* 84 (2009): 1079–1094.

[10] Genelex DNA Drug Sensitivity Test (DST) Results. Cytochrome P450CYP2C9 and VKORC1 (Warfarin, Coumadin) Test. http://www.healthanddna.com/SampleWarfarinReport.pdf.

[11] FAQs: Questions and Answers onHER2 Testing in Clinical Practice. Herceptin trastuzumab. http://www.herceptin.com/hcp/HER2-testing/faqs.jsp. Wikipedia. "HER2/neu." http://en.wikipedia.org/wiki/HER2/neu. Le, Xiao-Feng, Pruefer, and Robert C. Bast, Jr. "HER2-Targeting Antibodies Modulate the Cyclin-Dependent Kinase Inhibitor $p27^{Kip1}$ via Multiple Signaling Pathways." *Cell Cycle* 4 (2005): 87–95.

[12] Carter, Suzanne M., and Susan J. Gross. *e*medicine. "Glucose-6-Phosphate Dehydrogenase Deficiency." http://emedicine.medscape.com/article/200390-overview. "*ELITEK (rasburicase)* Mechanism of Action." http://www.elitekinfo.com/elitek_efficacy/MOA.aspx. "Hyperuricemia." http://www.elitekinfo.com/hyperuricemia/default.aspx. "Tumor Lysis Syndrome (TLS)." http://www.elitekinfo.com/tumor_lysis_syndrome/default.aspx.

[13] Watson, J. D., and F. H. C. Crick. "A Structure for Deoxyribose Nucleic Acid." *Nature* 171 (1953): 737–738.

[14] Avery, Oswald T., Colin M. MacLeod, and Maclyn McCarty. "Studies of the Chemical Nature of the Substance Inducing Transformation of Pneumococcal Types: Induction of Transformation by a Desoxyribonucleic Acid Fraction Isolated from Pneumococcus Type III." *Journal of Experimental Medicine* 79 (1944): 137–158.

[15] Hershey, A. D., and M. Chase. "Independent Functions of Viral Protein and Nucleic Acid in Growth of Bacteriophage." *Journal of General Physiology* 36 (1952): 39–56.

[16] Chargaff, E. "Chemical Specificity of Nucleic Acids and Mechanism of Their Enzymatic Degradation." *Experientia* 6 (1950): 201–209.

[17] For an excellent summary of the history of the Human Genome Project from the start of DNA sequencing through 2001, see Roberts, Leslie. "Controversial From the Start." *Science* 291 (2001): 1182–1188. And an overlapping article that carries the genome through its completion is Collins, Francis S., Michael Morgan, and Aristides Patrinos. "The Human Genome Project: Lessons from Large-Scale Biology." *Science* 300 (2003): 286–290.

[18] Wadham, Meredith. "James Watson's Genome Sequenced at High Speed." *Nature* 452 (2008): 788. Wheeler, David A. et al. "The Complete Genome of an Individual by Massively Parallel DNA Sequencing." *Nature* 452 (2008): 872–876.

[19] Davies, Kevin. "Illumina to Offer Personal Genome Sequencing Service." June 11, 2009. Bio.ITWorld.com. http://www.bio-itworld.com/BioIT_Article.aspx?id=92480&terms=June+11. Davies, Kevin. "Illumina Drops Personal Genome Sequencing Price to Below $20,000." Bio.ITWorld.com. http://www.bio-itworld.com/news/06/03/10/Illumina-personal-genome-sequencing-price-drop.html.

[20] Davies, Kevin, Maark Gabrenya, and Allison Proffit. "The Road to the $1,000 Genome." September 28, 2010. Bio-IT World.com. http://www.bio-itworld.com/2010/09/28/1Kgenome.html.

[21] Vorhaus, Dan. "Another Step on the Road to the $1,000 Genome." Genomics Law Report, January 12, 2010. http://www.genomicslawreport.com/index.php/2010/01/12/another-stop-on-the-road-to-the-1000-genome/.

[22] Drmanac, Radoje et al. "Human Genome Sequencing Using Unchained Base Reads on Self-Assembling DNA Nanoarrays." *Science* 327 (2010): 78–81.

[23] Tucker, Tracy, Marco Marra, and Jan M. Friedman. "Massively Parallel Sequencing: The Next Big Thing in Genetic Medicine." *The American Journal of Human Genetics* 85 (2009): 142–154.

[24] Davies, Kevin. "Splash Down: Pacific Biosciences Unveils Third-Generation Sequencing Machine." Bio.ITWorld.com. http://www.bio-itworld.com/2010/02/26/pacbio.html.

[25] Archon Genomics X Prize. http://genomics.xprize.org/archon-x-prize-for-genomics/prize-overview. Duncan, David Ewing. "On a Mission to Sequence the Genomes of 100,000 People." *The New York Times*, June 7, 2010.

[26] Davies, Kevin. *The $1000 Genome: The Revolution in DNA Sequencing and the New Era of Personalized Medicine*. New York: Free Press, 2010, pp. 174–179.

[27] Ibid. pp. 1–2.

[28] 23andme. "Breast Cancer. Sample Report." https://www.23andme.com/health/Breast-Cancer/.

[29] Davies, op. cit., 26, pp. 206–211. Knome, homepage. http://www.knome.com/index.html.

[30] See the excellent discussion in Collins, Francis. *The Language of Life: DNA and the Revolution in Personalized Medicine*. New York: Harper-Collins Publishers, 2010, pp. 79–81.

[31] Publication Announcement: Comparison of State Laws for Direct-to-Consumer Testing. Genetics & Public Policy Center. http://www.dnapolicy.org/news.release.php?action=detail&pressrelease_id=81.

[32] American College of Medical Genetics. "ACMG Statement on Direct-to-Consumer Genetic Testing." http://www.acmg.net/StaticContent/StaticPages/DTC_Statement.pdf. "ASHG Statement on Direct-to-Consumer Genetic Testing in the United States." *The American Journal of Human Genetics* 81 (2007): 635–637.

[33] "Evaluating the NIH's New Genetic Testing Registry." http://www.genomicslawreport.com/index.php/2010/03/18/evaluating-the-nihs-new-genetic-testing-registry/.

[34] National Human Genome Research Institute. "Genetic Information Nondiscrimination Act (GINA) of 2008." http://www.genome.gov/24519851. "GINA" The Genetic Information Nondiscrimination Act of 2008; Information for Researchers and Health Care Professionals. April 6, 2009. http://www.genome.gov/Pages/PolicyEthics/GeneticDiscrimination/GINAInfoDoc.pdf.

Glossary

allele

An alternative form of a gene (one member of a pair) that is located at a specific position on a specific chromosome. For example, the color of the pea seeds Mendel worked with was yellow or green. Hence, yellow and green are alleles at the same genetic locus. Sometimes there are more than two alleles at a given locus (multiple alleles).

amniocentesis

A prenatal test used to determine whether a fetus has a genetic defect. Following ultrasound to determine a safe place to insert a needle into the amniotic sac to withdraw amniotic fluid, a sample of amniotic fluid is collected. The whole procedure takes about 45 minutes. The amniotic fluid, containing fetal cells, is then subjected to laboratory analysis for chromosomal or genetic defects. Amniocentesis is usually performed between 14 and 20 weeks.

antisense

Antisense RNA is complementary to sense or messenger RNA. By binding to this RNA, antisense RNA renders it nonfunctional. In antisense therapy, the whole idea is to block the expression of a specific mRNA or mRNAs that encode proteins whose synthesis is deleterious.

candidate gene

Any gene thought to cause a disease. In the candidate gene method, people with and without the disease are compared to see whether there are any single nucleotide polymorphisms (see *single nucleotide polymorphism*) within the gene or in surrounding regions that differ between the two groups that might be related to the disease.

carcinoma

An invasive malignant tumor derived from epithelial cells.

chorionic villus sampling

A prenatal genetic test in which a sample of chorionic villi is removed from the placenta for testing. During pregnancy, the placenta provides oxygen and nutrients to the fetus and removes waste products from its blood. The chorionic villi are wispy projections that make up most of the placenta and include fetal tissue. Chorionic villus sampling usually occurs 10 to 13 weeks after the last menstrual period.

chromosome

DNA molecules in the nucleus of each cell are organized in thread-like structures called chromosomes. Each chromosome consists of a DNA molecule tightly coiled many times around proteins called histones that organize its structure. Chromosomes cannot be visualized in the nuclei of undividing cells, but they become visible under the microscope at the time of cell division because they become much more tightly packed. Specific stains are usually employed that make the chromosomes easier to see.

Each chromosome has a constriction point called the centromere, which divides the chromosome into two parts, or "arms." The short arm of the chromosome is labeled the "p arm." The long arm of the chromosome is labeled the "q arm." Chromosomes can be distinguished on the basis of size and the relative positions of their centromeres.

code

The genetic code consists of codons of three letters each. There are 64 codons in the genetic code, but they specify only 20 amino acids. As a consequence, the code is said to be degenerate, meaning that more than one codon usually specifies a given amino acid. For example, the amino acid serine is specified by the codons UCU, UCC, UCA, UCG, AGU, and AGC, whereas proline has the codons CCU, CCC, CCA, and CCG. Note that codons are usually specified by their RNA letters, so U substitutes for the letter T in DNA. There is no internal punctuation in the genetic code. However, the codon at the beginning of the genetic sentence is usually AUG, which specifies methionine, whereas the three termination or stop codons at the end of the sentence are UAA, UAG, and UGA.

diploid

Diploid refers to the fact that somatic cells in animals and plants usually have two sets of chromosomes (2n) as opposed to haploid eggs and sperm that are haploid (n), having one set of chromosomes.

diseasome

A network of disorders and disease genes linked by known disorder-gene associations.

dominant

A dominant gene is the one whose expression one sees even if its recessive allele is present on the homologous chromosome. For example, yellow seed color in Mendel's peas (Y) is dominant over green (y), which is recessive. So when plants that form yellow seeds (YY) are crossed with plants whose seeds are green (yy), the progeny are all yellow even though they are genotypically (Yy).

dosage compensation

In humans and other mammals, dosage compensation usually refers to the fact that, although the female sex usually has two sex (XX) chromosomes and the male has just one plus a Y chromosome (XY), only one X chromosome in the female is active, while its homolog is inactivated. However, a female is a mosaic as the paternal X chromosome is active in some tissues while the maternal X chromosome is active in other tissues.

egg

The term *egg* tends to be used rather loosely by fertility clinics, most of which have egg donation programs and offer eggs for fertilization to infertile couples. In fact, what they are offering are secondary oocytes that have not completed the second division of meiosis. Upon fertilization, these secondary oocytes complete the second division of meiosis yielding a fertilized egg. So humans do not produce unfertilized eggs. See *oocyte and oogenesis* for a more complete description.

epigenetic

An inherited alteration in gene function that occurs without altering the DNA sequence. For example, methylation of a gene sequence may alter or block the gene from functioning. Another example is the inactivation of one X chromosome out of the two possessed by a female. See *dosage compensation*.

epigenome
The heritable factors such as methylation of nucleotides that modify genomic expression without altering DNA sequence.

exon
The genes of humans and other higher organisms are split into coding sequences called exons and noncoding sequences called introns. The introns are spliced out of the messenger RNA following transcription so that the exons form a continuous sequence for translation.

FISH
Fluorescence *in situ* hybridization is a technique that is used to detect and localize the presence or absence of specific chromosomal DNA sequences. FISH uses fluorescent probes that bind to only those parts of the chromosome with which they possess a high degree of sequence similarity.

gene
The unit of heredity. Genes are DNA sequences that usually encode specific proteins or polypeptides. Most genes of higher organisms, including humans, are split into coding sequences (exons) and noncoding sequences (introns). See *exon*.

genome
The sum total of an organism's genetic information.

genome-wide association study (GWAS)
GWAS compares the genomes of people with the disease or condition of interest to those who lack it to search for specific genetic markers that are associated with the disease or condition. See *indel, single nucleotide polymorphism.*

genotype
The genetic constitution of an organism as compared with the phenotype or appearance of the organism. For example, a pea plant that has yellow seeds (phenotype) may have the genotype YY, meaning it is pure (homozygous) for the Y gene, or it may have the genotype (Yy), meaning it is a carrier (heterozygous) for the green gene.

haploid
See *diploid.*

haplotype

A contraction of haploid genotype. A haplotype is a set of tightly linked genetic markers present on a chromosome that are usually inherited as a group and not separated by recombination.

HapMap

A haplotype map of the human genome. See *haplotype*.

heterozygote

See *genotype*.

homozygote

See *genotype*.

Human Genome Project

The international research effort to sequence and map all of the human genes, collectively known as the genome. The project attained its goal in 2003 with the first detailed map of the human genome.

immune system

A system within an organism that protects the organism by identifying and destroying pathogens and tumor cells. The immune system detects a wide variety of undesirable agents, from viruses to parasitic worms. To function properly, the immune system has to distinguish these foreign agents (nonself) from the organism's own healthy cells and tissues (self). The "adaptive immune system" is responsible for a strong immune response to a foreign agent. This system also functions to maintain immunological memory in which each pathogen is "remembered" by a signature antigen. The principal cellular workhorses of this system are B and T lymphocytes.

B and T cells possess receptor molecules that recognize specific targets. T cells recognize a "nonself" target like a pathogen, only after fragments of the pathogen (antigens) have been processed and presented in combination with a "self" receptor called a major histocompatibility complex (MHC) molecule (see *MHC locus*). There are two major varieties of T cells: the killer T cell and the helper T cell. Killer T cells only recognize antigens coupled to Class I MHC molecules, whereas helper T cells only recognize antigens coupled to Class II MHC molecules. A B cell recognizes foreign antigens such as those produced by pathogens when the antigen binds to an antibody on its cell surface. The antigenic molecules are then displayed on MHC Class II molecules on the surface of the B cell. The MHC-antigen combination then attracts a matching helper T cell. The T cell releases lymphokines. These activate the B cell, causing it to divide with the daughter cells secreting millions of copies of the antibody molecule. These antibodies then circulate in the blood stream, following which they bind to the pathogens marking them for destruction.

immunoglobulins

Proteins synthesized by plasma cells and lymphocytes that play a crucial role in the immune system. These proteins attach to foreign substances, such as bacteria, and help in their destruction. The classes of immunoglobulins are named immunoglobulin A (IgA), immunoglobulin G (IgG), immunoglobulin M (IgM), immunoglobulin D (IgD), and immunoglobulin E (IgE).

indel

Insertion or deletion mutations.

intron

See *exon*.

in vitro fertilization (IVF)

A method that permits sperm to fertilize eggs outside of the body.

LINEs and SINEs

Long and short interspersed nuclear elements, respectively. They invade new genomic sites using RNA intermediates. These elements are found in almost all higher organisms and together account for at least 34% of the human genome. LINEs move from one location to another by a copy-and-paste mechanism. Copying of LINE DNA into RNA is accomplished by the cellular RNA polymerase II that also transcribes genes into mRNA. A reverse transcriptase encoded by the LINE then copies this RNA into DNA, following which is it inserted into a specific target sequence in the DNA following a cut made at that site by the endonuclease encoded by the LINE. The noncoding SINEs depend on reverse transcriptase and endonuclease functions encoded by partner LINEs.

linkage

The tendency of certain genes to stay together. Thus, genes that are physically close together on a chromosome tend to stay together most of the time, whereas those that are further apart tend to recombine with each other. Genetic maps can be constructed for linked genetic markers where the measure of distance between any two genes is represented by the frequency with which they recombine. See also *meiosis*.

linkage disequilibrium

The occurrence in a population of certain combinations of linked alleles in greater proportion than expected from the allele frequencies at the loci.

lymphocyte

See *immune system*.

meiosis

The cellular division process that leads to the production of sperm and eggs while reducing the chromosome number from the diploid (2n) to the haploid (n) level. It consists of two divisions preceded by a round of DNA replication. Pairing of homologous paternal and maternal chromosomes occurs at Metaphase of meiosis I.

Let's consider an organism with two pairs of chromosomes. The first pair is marked by gene A on the paternal chromosome and its allele a on the maternal chromosome. The second pair is marked gene B on the paternal chromosome and its allele b on the maternal chromosome. These chromosome pairs will assort independently at the first meiotic division and then segregate into sperm and eggs at the second meiotic division yielding AB, Ab, aB, and ab sperm or eggs in equal frequency. Note that, if the first chromosome pair has an additional pair of genetic alleles present—let's call them C on the paternal chromosome and c on the maternal chromosome—they can also undergo recombination. So, looking just at the A and C genes, we will have AC, Ac, aC, and ac sperm and eggs, but they will not be in equal frequency unless the two genes are far apart on the chromosome. If they are very close together, parental AC and ac sperm and eggs will be at a much higher frequency than recombinant Ac and aC sperm and eggs. The frequency of the recombinants can be used to obtain a genetic map distance between the two genes. So, if 10% of the sperm or eggs are recombinant, we say they are 10 map units apart.

Mendelian genetics

Mendel's first law relates to the segregation of characters. Suppose you cross pea plants that produce yellow seeds (YY) with plants whose seeds are green (yy). The progeny will be yellow phenotypically, but genotypically they will be Yy. If we now cross these F1 progeny to each other, the genotypic ratio of the progeny in the next generation (F2) will be 1 YY: 2YY: 1yy, but because yellow is dominant to green, the phenotypic ratio will be 3 yellow plants: 1 green plant.

Mendel's second law relates to the independent assortment of different characters. Now let's bring in a second character affecting seeds. Sometimes they are smooth (SS) and sometimes wrinkled (ss). If we cross plants that have smooth seeds with wrinkled-seeded plants, we get plants with smooth seeds, so smooth must be dominant to wrinkled. If we now cross yellow and smooth-seeded plants (YYSS) with green and wrinkle-seeded plants (yyss), we get plants that produce yellow, smooth seeds, but these F1 plants are genotypically YySs. If we cross these F1 plants with each other in the next generation (F2), we will get a 9 yellow smooth (Y_S_): to 3 yellow wrinkled (Y_ss): 3 green smooth (yyS_): 1 green wrinkled (yyss). In other words, the two genes assort independently. Note that a _ means the remaining allele is either Y or y or S or s. You can verify the 9:3:3:1 ratio for yourself by constructing a Punnett square.

MHC locus

The major histocompatibility complex (MHC) is the most gene-dense region of the mammalian genome and includes the many genes that encode MHC molecules. These molecules play an important role in the immune system. When the MHC-protein complex, displayed on the surface of a cell, is presented to an immune system cell such as a T cell or natural killer (NK) cell, the immune cell can kill the infected cell, and other infected cells displaying the same proteins if they are recognized as nonself. Class I MHC molecules are found on most cells and present foreign proteins to killer T cells. Class II MHC molecules are found on certain immune cells themselves, mainly macrophages and B-lymphocytes, also known as antigen-presenting cells. These antigen-presenting cells ingest microbes, destroy them, and digest them into fragments. The Class II MHC molecules on the antigen-presenting cells present the fragments to helper T cells that stimulate other cells to elicit an immune reaction.

mitosis

The process by which the chromosomes are separated into two identical packages at the time of cell division.

monoclonal antibody

A pure antibody directed against a specific target.

mutation types

A *missense* mutation in a gene replaces one base pair (e.g., A-T) with a second (e.g., G-C). The resulting change causes an amino acid substitution in the protein encoded by the gene, rendering it nonfunctional or poorly functional. A *neutral* mutation is like a missense mutation except that the amino acid change does not affect the function of the protein. A *silent* mutation is like a neutral mutation except the codon change does not result in an amino acid change. *Nonsense* mutations replace a codon specifying an amino acid in protein with UAG, UGA, or UAA. Each of these codons signals the protein synthesizing apparatus to stop at that point. This results in polypeptide chain termination and an incomplete protein. *Frameshift* mutations insert or remove a base pair. This causes a change in the reading frame of the messenger RNA so that amino acids not originally specified to be included in the protein are added to the protein. Frequently, a newly created premature stop codon is encountered, resulting in the abrupt termination of the incomplete polypeptide chain.

nucleotide

A nucleotide is composed of a nitrogenous base usually one of two purines, adenine (A) or guanine, or one of two pyrimidines, cytosine (C) and either thymine (T) in DNA or its equivalent uracil (U) in RNA, a five-carbon sugar, either ribose in RNA or 2'-deoxyribose in DNA, and one to three phosphate groups.

oncogene

A gene that when mutated or expressed at high levels when it should not be is instrumental in turning a normal cell into a cancer cell.

oocyte and oogenesis

The ovary contains many follicles composed of a developing egg surrounded by an outer layer of follicle cells. Each egg begins oogenesis as a primary oocyte. At birth, each female carries a lifetime supply of primary oocytes, each of which is arrested early in the first meiotic division (Prophase I). Following completion of the first meiotic division, a secondary oocyte is released each month from puberty until menopause, a total of 400–500. If a secondary oocyte encounters a sperm, it completes the second meiotic division to yield a fertilized egg. At each meiotic division, a polar body cell is also formed that eventually disintegrates.

phenotype

Any observable characteristic or trait of an organism. The phenotype does not necessarily reflect the genotype. For example, pea plants with yellow seeds may either be homozygous (YY) or heterozygous for the green allele that is recessive (Yy) to the dominant yellow allele.

polymerase chain reaction (PCR)

PCR permits amplification of single or a few copies of a piece of DNA to yield millions of copies of that DNA sequence. The method relies on repeated heating and cooling cycles. The heating cycle melts the hydrogen bonds of the DNA helix, yielding single strands of DNA. Short DNA sequences (primers) complementary to bits of the target region are added to the mix. This permits a heat-stable DNA polymerase from a bacterium called *Thermus aquaticus* (Taq) to recognize the primers and replicate the DNA of the target sequence at each cycle. As PCR progresses, the DNA generated at each cycle serves as a template for replication, setting in motion the exponential amplification of the target DNA.

polymorphism

Refers to the presence of one of two or more variants of a particular DNA sequence in a population. Single base pair polymorphisms are the most common, but polymorphisms are sometimes much larger in size.

polypeptide

A peptide is made up of at least two amino acids. Longer chains of amino acids are referred to as polypeptides. Many proteins are made up of a single type of polypeptide, but others—hemoglobin is an example—possess more than one distinct polypeptide chain.

positional cloning

A method employed to locate a disease-associated gene on its chromosome. This method works even if there is a paucity of information about the molecular basis of the disease. Genetic linkage analysis is used to identify molecular markers close to the gene. Once the approximate location of the gene has been determined, partially overlapping DNA segments are isolated that progress along the chromosome until they include the candidate gene. This is how Francis Collins and colleagues isolated the cystic fibrosis gene.

preimplantation genetic diagnosis (PGD)
A method used to determine whether embryos created by in vitro fertilization have genetic defects before the embryos are implanted. For example, if a couple wants to have children and both are carriers of a cystic fibrosis gene mutation, one in four of the embryos would be expected to be homozygous for the mutation, meaning they will have the disease. PGD would be used to identify these embryos so they would not be implanted.

recessive
See *dominant*.

recombinant gene or protein
A gene or genes that have been spliced into the DNA of another species. The introduced gene becomes part of the genetic apparatus of the host and programs the synthesis of a recombinant protein. The proteins produced are often ones that have important therapeutic uses.

recombination
Genetic recombination is a process by which a molecule of DNA, but occasionally RNA, is broken and then joined to a different one. Recombination can occur between similar molecules of DNA (homologous recombination) or dissimilar molecules (nonhomologous recombination). In higher organisms, recombination, or crossing over, between homologous chromosomes occurs during the first meiotic division. The crossing over process recombines genes on homologous chromosomes in different combinations. See also *meiosis*.

repair
DNA repair includes several different processes by which defects in the DNA, often induced by UV radiation or chemical mutagens, are identified and corrected. UV light induces the formation of thymine dimers, which are corrected by the photolyase enzyme that requires visible light to function. There are three other main "dark repair" (i.e., light is not required for function) repair systems. One of these excises damaged bases, the second removes damaged nucleotides, and the third corrects mismatches in DNA (e.g., A-C or G-T).

retrotransposon
See *LINEs and SINEs*.

retrovirus

An RNA virus that encodes a protein called reverse transcriptase. This enzyme catalyzes the synthesis of a DNA copy from RNA genome whose incorporation into the genome of the host cell is facilitated by an integrase enzyme also encoded in the viral genome.

sarcoma

A cancer that arises from connective tissue cells. These cells originate from embryonic mesoderm, or the middle layer that forms the bone, cartilage, and fat tissues.

sex-linked

Genes on either the X or Y chromosome are said to be sex-linked.

single nucleotide polymorphism (SNP)

DNA sequence variations that occur when a single nucleotide (A, T, C, or G) in the genome sequence is altered. For example, an SNP might change the DNA sequence AAGGCTAA to ATGGCTAA. For a variation to be classified as an SNP, it must be present in at least 1% of the population. SNPs comprise about 90% of all human genetic variation. SNPs are sometimes found in genes, but are often found in noncoding sequences of the human genome. Most SNPs probably have little effect, but some may predispose a person to a certain genetic disease.

transcription

The first step in gene expression. It involves making an RNA copy of one strand of the DNA under the aegis of the enzyme RNA polymerase. This genetic message (mRNA) can then be translated into protein once the introns have been spliced out.

transgene

A gene that has been transferred from one organism to another either naturally or through the use of genetic engineering.

translation

The process by which the information contained in a messenger RNA molecule is converted into the amino acid sequence of the polypeptide encoded by that molecule.

translocation

A chromosome alteration in which a whole chromosome or segment of a chromosome becomes attached to or interchanged with a nonhomologous chromosome or chromosome segment.

transposon

DNA transposons in contrast to retrotransposons (see *LINEs and SINEs*) do not make use of an RNA intermediate in the transposition process. A transposase enzyme catalyzes the transposition process. Some transposases can bind nonspecifically to any target site, whereas others bind to specific sequence targets.

tumor suppressor

A gene that reduces the probability that a normal cell will be converted to a cancer cell. For example, the products of certain tumor suppressor genes spot mutational mistakes that occur during DNA replication and slow down the process so that they can be repaired. When this mechanism fails, some of those mutations will cause the cell to become cancerous, for example, when an oncogene is activated as the result of mutation.

wild type

The term geneticists use to describe the typical form of a species as it occurs in nature. For example, wild type fruit flies have red eyes, but there is a host of mutants that affect eye color.

Some useful human genetics Web sites

All about the Human Genome Project (HGP): http://www.genome.gov/10001772

GeneReviews: http://www.ncbi.nlm.nih.gov/sites/GeneTests/?db=GeneTests

Genetics Home Reference: http://ghr.nlm.nih.gov/

Human Genome Resources: http://www.ncbi.nlm.nih.gov/genome/guide/human/

Online Mendelian Inheritance in Man (OMIM): http://www.ncbi.nlm.nih.gov/omim

Acknowledgments

I want to acknowledge with pleasure my friends Ned Arnett, Joe Blum, Clark Havighurst, Sy Mauskopf, and Bob Wilkins who read and commented on some bits and pieces of this book at a very early stage in its writing. My longtime editor and friend Kirk Jensen has provided invaluable guidance as always. And my wife Carol has put up with my efforts that quite often got in the way of other things I was supposed to do with such good humor and understanding. I, of course, take full responsibility for any errors that have crept into the book.

About the author

Nicholas Wright Gillham is James B. Duke Professor of Biology Emeritus. His research interests involved the genetics and molecular biology of cellular organelles called chloroplasts and mitochondria. For more than a decade he taught a course entitled "The Social Implications of Genetics." This course fostered his interest in eugenics, human genetics, and their history. He has authored two books on chloroplasts and mitochondria plus a biography of the Victorian scientist Francis Galton entitled, *A Life of Sir Francis Galton: From African Exploration to the Birth of Eugenics* (Oxford University Press, 2001).

Index

Numbers

A

B

C

D

E

F

G

H

I

J

M

N

O

P

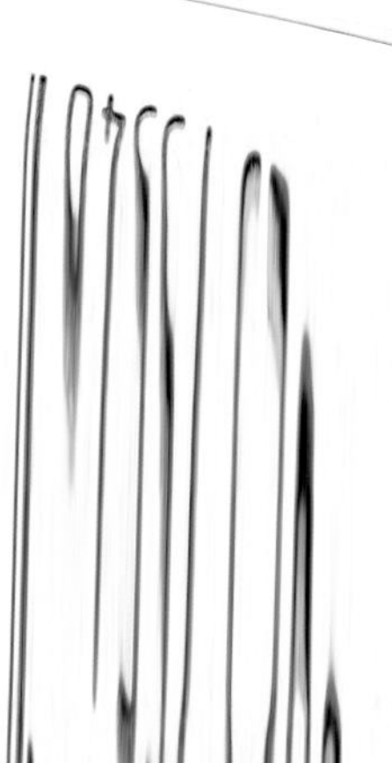

Q–R

S

X–Y–Z

The life sciences revolution is transforming
our world as profoundly as the industrial
and information revolutions did
in the last two centuries.
FT Press Science will capture the excitement and
promise of the new life sciences, bringing breakthrough
knowledge to every professional and interested citizen.
We will publish tomorrow's indispensable work in
genetics, evolution, neuroscience, medicine,
biotech, environmental science, and whatever
new fields emerge next.
We hope to help you make sense of the future,
so you can *live* it, *profit* from it, and *lead* it.